T0213999

Studies in Computational Intelligence

Volume 977

Series Editor

Janusz Kacprzyk, Polish Academy of Sciences, Warsaw, Poland

The series "Studies in Computational Intelligence" (SCI) publishes new developments and advances in the various areas of computational intelligence—quickly and with a high quality. The intent is to cover the theory, applications, and design methods of computational intelligence, as embedded in the fields of engineering, computer science, physics and life sciences, as well as the methodologies behind them. The series contains monographs, lecture notes and edited volumes in computational intelligence spanning the areas of neural networks, connectionist systems, genetic algorithms, evolutionary computation, artificial intelligence, cellular automata, self-organizing systems, soft computing, fuzzy systems, and hybrid intelligent systems. Of particular value to both the contributors and the readership are the short publication timeframe and the world-wide distribution, which enable both wide and rapid dissemination of research output.

Indexed by SCOPUS, DBLP, WTI Frankfurt eG, zbMATH, SCImago.

All books published in the series are submitted for consideration in Web of Science.

More information about this series at http://www.springer.com/series/7092

Stefan Kollmannsberger · Davide D'Angella ·
Moritz Jokeit · Leon Herrmann

Deep Learning
in Computational Mechanics

An Introductory Course

Springer

Stefan Kollmannsberger
School of Engineering and Design
Chair of Computational Modeling
and Simulation
TU Munich, München, Germany

Davide D'Angella
School of Engineering and Design
Chair of Computational Modeling
and Simulation
TU Munich, München, Germany

Moritz Jokeit
Department of Health Sciences
and Technology
Institute for Biomechanics
ETH Zürich, Zürich, Switzerland

Leon Herrmann
Bavarian Graduate School
of Computational Engineering
TU Munich, München, Germany

ISSN 1860-949X   ISSN 1860-9503 (electronic)
Studies in Computational Intelligence
ISBN 978-3-030-76589-7   ISBN 978-3-030-76587-3 (eBook)
https://doi.org/10.1007/978-3-030-76587-3

This Springer imprint is published by the registered company Springer Nature Switzerland AG
The registered company address is: Gewerbestrasse 11, 6330 Cham, Switzerland

Contents

Chapter 1
Introduction

The pursuit of artificial intelligence dates back to the time computers in a modern sense were born [RND10]. Ever since, computer science has been partially dedicated to finding the key to intelligent machines. Over the last few years, rapid advancements in the field of machine learning have caused quite a sensation. Algorithms suddenly achieved competing results in tasks thought to be the reserve of human intelligence. Reasons for their success are the vast availability of data paired with a straightforward implementation and powerful graphics hardware. Nowadays, new classes of learning machines are confronted with various complex problems. Whether audiovisual signal processing, weather forecasting, or autonomous driving, data-driven algorithms offer tremendous potential and take computer science a step closer to the ideal of intelligent machines.

The unprecedented success of machine learning has not gone unheeded in other scientific fields. Especially in the presence of adequate data, the application of learning-based algorithms has shown promising results. For instance, image-based medical diagnosis already benefits from outstanding progress in visual object recognition [Lam+12]. At the same time, the demand for data constitutes a major obstacle for wider interdisciplinary adoption. Physics and engineering applications often lack sufficient information about the underlying problem. Thus, the use of data-driven approaches seems rather naive. Further, engineering sciences have an established, long-standing paradigm of computer-aided problem solving. Finding numerical solutions to problems that arise from observing natural or engineered systems describes the essence of computational mechanics. The solution process commences with formalizing the problem in terms of physical quantities. A set of partial differential equations typically governs the resulting mathematical description. Further, computational methods require a discretization of the problem into finite elements to allow the numerical approximation of continuous variables inside a prescribed domain. Most methods following this scheme have been developed and enhanced over decades, hence guaranteeing a high level of robustness and reliability.

© The Author(s), under exclusive license to Springer Nature Switzerland AG 2021
S. Kollmannsberger et al., *Deep Learning in Computational Mechanics*,
Studies in Computational Intelligence 977,
https://doi.org/10.1007/978-3-030-76587-3_1

Nevertheless, the generality and simplicity of learning algorithms have sparked interest in the scientific computing community and inspired early attempts toward the data-driven solution of partial differential equations. Incorporating domain knowledge into the learning model allowed it to compensate for the typical data sparsity in physical problems. The computational limitations at that time restricted the application beyond the scope of canonical examples. However, the availability of enhanced programming environments and potent hardware has encouraged researchers from various disciplines to revisit the proposed ideas over the past years. Current publications demonstrate the potential of physics-enriched surrogate models for the inference and identification of partial differential equations [RPK19, Sam+19, NM19]. Most approaches deploy neural network architectures to approximate the hidden solution. Neural networks loosely mimic the neuronal structure of the brain and belong to the most successful group of contemporary learning algorithms. Their ability to learn highly non-linear representations is an excellent prerequisite for modeling physical phenomena.

This manuscript is not an expert book on how to design neural networks. Other books, such as [GBC16] are a much better resource for this topic.

Instead, this manuscript is a slim and concise introduction into the topic of how learning based methods in general and neural networks in particular are used to solve challenges in the field of computational mechanics. To this end, it teaches you the absolute essence of what is needed to solve the simplest possible examples such as well known 1D bar problems. Of course, no one would use the presented methodologies to solve such simple tasks. But the core concepts taught by this manuscript generalize to address more complex problems. And working through this manuscript will put you in the position to do just that: to start your own endeavor into the emerging field of deep learning in computational mechanics.

Following this introduction, Chap. 2 leads off with the fundamental concepts of machine learning. Subsequently, Chap. 3 elaborates on neural networks and their corresponding algorithms. After reviewing machine learning applications in engineering and physics, Chap. 4 shifts the focus toward physics-enriched deep learning models. Chapter 5 documents the implementation of a physics-informed neural network on a static bar and a transient heat transfer example. The chapter concludes with a discussion of results and future research directions. Finally, Chap. 6 introduces a neural network using a variational approach called the deep energy method, which is also applied to a static bar example.

References

[RND10]   Stuart J. Russell, Peter Norvig, and Ernest Davis. *Artificial intelligence: a modern approach*. 3rd ed. Prentice Hall series in artificial intelligence. Upper Saddle River: Prentice Hall, 2010. 1132 pp. ISBN: 978-0-13-604259-4.

[Lam+12] Philippe Lambin et al. "Radiomics: Extracting more information from medical images using advanced feature analysis". In: *European Journal of Cancer* 48.4 (Mar. 2012), pp. 441–446. ISSN: 09598049. DOI https://doi.org/10.1016/j.ejca.2011.11.036. URL: https://linkinghub.elsevier.com/retrieve/pii/S0959804911009993 (visited on 07/02/2020).

[RPK19] M. Raissi, P. Perdikaris, and G.E. Karniadakis. "Physics-informed neural networks: A deep learning framework for solving forward and inverse problems involving nonlinear partial di erential equations". In: *Journal of Computational Physics* 378 (Feb. 2019), pp. 686–707. ISSN: 00219991. DOI https://doi.org/10.1016/j.jcp.2018.10.045. URL: https://linkinghub.elsevier.com/retrieve/pii/S0021999118307125 (visited on 01/08/2020).

[Sam+19] Esteban Samaniego et al. "An Energy Approach to the Solution of Partial Differential Equations in Computational Mechanics via Machine Learning: Concepts, Implementation and Applications". In: arXiv:1908.10407 [cs, math, stat] (Sept. 2, 2019). URL: http://arxiv.org/abs/1908.10407 (visited on 01/08/2020).

[NM19] Mohammad Amin Nabian and Hadi Meidani. "A Deep Neural Network Surrogate for High-Dimensional Random Partial Differential Equations". In: *Probabilistic Engineering Mechanics* 57 (July 2019), pp. 14–25. ISSN: 02668920. https://doi.org/10.1016/j.probengmech.2019.05.001. URL: http://arxiv.org/abs/1806.02957 (visited on 02/21/2020).

[GBC16] Ian Goodfellow, Yoshua Bengio, and Aaron Courville. *Deep Learning*. MIT Press, 2016. ISBN: 0-262-03561-8. URL: http://www.deeplearningbook.org.

Chapter 2
Fundamental Concepts of Machine Learning

2.1 Definition

Nowadays, machine learning is arguably the most successful and widely used technique to address problems that cannot be solved by hand crafted programs. In contrast to conventional algorithms following a predefined set of rules, a machine learning algorithm relies on a large amount of data that is observed in nature, handcrafted by humans, or generated by another algorithm [Bur19]. A more formal definition by Mitchell states that "a computer program is said to learn from experience E with respect to some class of tasks T and performance measure P, if its performance at tasks in T, as measured by P, improves with experience E" [Mit97]. Taking image recognition as an example, the task T is to classify previously unseen images, the performance measure P corresponds to the amount of correctly classified images, and the experience E includes all images that have been used to train the algorithm. Most machine learning algorithms can be decomposed into the following features: a dataset, a cost function, an optimization procedure, and a parameterized model [GBC16]. Generally, the cost function defines an optimization criterion by relating the data to the model parameters. Further, the optimization procedure searches for the model parameters representing the provided data best. The key difference between machine learning and solving an optimization problem is that the optimized model is then used for predictions on previously unseen data.

Electronic supplementary material The online version of this chapter (https://doi.org/10.1007/978-3-030-76587-3_2) contains supplementary material, which is available to authorized users.

2.2   Data Structure

A machine learning algorithm processes a dataset containing a collection of data points often referred to examples. Every example consists of one or more features describing the data point in a quantitative manner. In terms of notation, each example can be written as a vector x, where each entry x_j corresponds to a feature of that example. To take advantage of fast implementations of matrix vector calculus in modern programming languages, all examples are arranged in a so-called design matrix

$$
X \quad = \quad
\begin{array}{c}
\text{example 1} \\
\text{example 2} \\
\vdots \\
\text{example } m
\end{array}
\overset{\displaystyle \text{feature 1}\quad \text{feature 2}\quad \cdots \quad \text{feature } n}{
\begin{bmatrix}
x_{11} & x_{12} & \cdots & x_{1n} \\
x_{21} & x_{22} & \cdots & x_{2n} \\
\vdots & \vdots & \ddots & \vdots \\
x_{m1} & x_{m2} & \cdots & x_{mn}
\end{bmatrix}}.
$$

Each row of the matrix represents an example and each column corresponds to a feature describing the examples [GBC16]. In the case of recognizing gray-scale images, every photo in the dataset is stored as one example vector in the design matrix. Assuming all pictures have a resolution of 50×50 pixels, then every example consists of 2500 features storing the gray-scale value for each pixel.

It is common practice to split the data into different subsets, namely a training set and a test set. The majority of the data, e.g. $\sim$90%, is included in the training set and used for learning the optimal parameters of the model. The remaining part, e.g. $\sim$10%, is kept for the test set to get an estimate for the model's performance on unseen data. The given percentages are for guidance only, since machine learning literature does not prescribe specific figures and leaves it to the practitioner on how to subdivide the data [Ng20].

2.3   Types of Learning

There exist different types of learning, which are briefly presented in the following subsections.

2.3.1   Supervised Learning

Most problems solved by machine learning algorithms fall into the category of supervised learning [Cho18]. In this context, "supervised" indicates that the algorithm is processing a labeled dataset. Thus, next to the design matrix, the dataset comprises a vector **y** with a label or target y_i for each example. For instance, in image classification tasks, each image has previously been annotated receiving a certain category label. The supervised learning algorithm studies the dataset and learns to classify the images into the given categories by comparing its prediction with the given ground truth label.

2.3.2   Unsupervised Learning

The goal of unsupervised learning is to find a structure or, more precisely, the probability distribution in the provided data. The data is not labeled and therefore no explicit prediction is possible. However, it can be very useful to apply unsupervised learning to large datasets in order to find inherent structures or repeating patterns in the data. For instance, anomaly detection algorithms are used to identify fraudulent credit card transactions that differ from the usual purchasing behavior of a customer [GBC16].

2.3.3   Semi-supervised Learning

As the term suggests, semi-supervised learning combines the two preceding concepts. In cases where only small samples of the data are labeled, unsupervised learning helps to improve the performance of the supervised learning algorithm [GBC16].

2.3.4   Reinforcement Learning

The basic idea of reinforcement learning is that an algorithm interacts with an environment to learn a certain decision behavior maximizing the expected average reward [Bur19]. It is used for problems involving sequential decision-making in order to fulfill a long-term goal. A group of Google researchers demonstrated the effectiveness of this technique on the example of playing the game of Go [Sil+17], an old and very complex board game that originated in China. Solely by playing games against itself, the program reached superhuman abilities and was capable of beating the European Go champion.

2.4   Machine Learning Tasks

The following subsections give a small overview of machine learning tasks with a short introduction to suitable algorithms. This list is by far not complete; however, many problems that arise in practice can be related to one of the following categories.

2.4.1 Regression

Regression is a supervised learning problem with the goal of predicting a numerical value. Basically, a regression algorithm outputs a function that maps a given input to an output, usually in form of a real number. An example is the prediction of house prices based on certain criteria like the area, number of rooms, or the age of the house. A more detailed explanation of linear regression can be found in Sect. 2.5. Further, it is possible to extend the algorithm for the use with polynomials as well as for the prediction of multiple outputs, known as multivariate regression. Decision tree algorithms and neural networks are also used for regression problems. The latter is introduced in Chap. 3.

2.4.2 Classification

Just like regression, classification tasks belong to the category of supervised learning. Instead of a numerical value, their output takes on a discrete form. In other words, a classification algorithm returns a function that assigns a category to the provided input. Again, the example of classifying the content of an image falls into this class of problems. Even though the name suggests otherwise, logistic regression is a classification algorithm generating a binary output based on the logistic (or sigmoid) function (cf. Chap. 3). Other important algorithms for categorization are support vector machines and decision trees like the random forest or the more advanced gradient boosting approach. In terms of performance at complex tasks like image recognition, those algorithms have been surpassed by neural networks, but are still used for simpler problems [GBC16].

2.4.3 Clustering

Clustering differs from classification and regression as it is an unsupervised learning task. The algorithm gives feedback about which parts of the data share similarities and therefore belong to the same cluster. A popular choice for clustering is the k-means algorithm that divides the incoming data into k different clusters of examples being close to each other [GBC16].

2.5 Example: Linear Regression

Even though linear regression is a very simple algorithm, it is well suited to explain concepts also applicable to more sophisticated machine learning algorithms. The

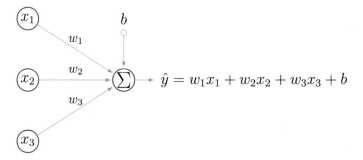

Fig. 2.1 Linear regression for a single example x with three input features

goal of a regression model is to predict a scalar value $\hat{y} \in \mathbb{R}$ from an input vector $x \in \mathbb{R}^n$. In general, linear regression can be written as

$$\hat{y} = w^T x + b = \sum_{j=1}^{n} w_j x_j + b, \tag{2.1}$$

where $\hat{y}$ represents the target, w is a vector containing the weights, and x denotes the example vector (cf. Fig. 2.1). The constant b is called bias, referring to the case when either the weights or the input is close to zero, the output is biased toward b. Each weight w_j is a coefficient that gets multiplied with the corresponding feature x_j. The magnitude and sign of a weight decide about the feature's importance to the prediction $\hat{y}$. For instance, a feature x_j has a corresponding weight w_j which is large in magnitude, then even small changes in x_j alter the prediction $\hat{y}$ by a great amount. The weights and the bias are the parameters of the model and determine how well it performs on the task of predicting target values for new data. In order to get an estimate of the model's future performance, a small part of the labeled data is held back to be used for evaluation. This fraction is called test set while the remaining part used for finding the optimal parameters is referred to as training set (cf. Sect. 2.2).

What is still missing is a way to measure the performance of the model. A common metric for this purpose is the squared error loss. If $\hat{y}$ is the prediction and y the ground truth, the squared error loss is simply defined as $(y - \hat{y})^2$. However, this only provides feedback for a single data point (x_i, y_i). To get one measurement for the performance on the whole dataset $\{(x_i, y_i)\}_{i=1}^{m}$, the mean of the squared error (MSE) of all examples m is calculated as follows:

$$MSE = \frac{1}{m} \sum_{i=1}^{m} (y_i - \hat{y}_i)^2. \tag{2.2}$$

The closer the prediction $\hat{y}_i$ gets to the value of y_i, the smaller the error and the better the model is expected to perform. The mean squared error is a popular choice for the cost function. It has a continuous derivative and naturally penalizes large differences between the true target and the prediction [Bur19]. In the case of linear regression, the squared error loss even leads to a convex optimization problem, meaning the cost function has only one particular minimum.

Now the question is how to find the optimal parameters w^* and b^* that yield a good prediction. This task can be interpreted as an optimization problem with the goal to minimize the mean squared error. Given Eq. (2.2) and inserting equation Eq. (2.1), a cost function $C(w, b)$ dependent on the parameters w and b is obtained

$$C(w, b) = \frac{1}{m} \sum_{i=1}^{m} (y_i - (w^T x_i + b))^2. \tag{2.3}$$

With the given cost function, it is possible to formulate the optimization problem: Find the model parameters w and b that minimize the cost function $C(w, b)$ or

$$\min_{w,b} C(w, b) = \min_{w,b} \frac{1}{m} \sum_{i=1}^{m} (y_i - (w^T x_i + b))^2. \tag{2.4}$$

Finding the optimal parameters that lead to a good performance of the model can be interpreted as "learning".

The general concept that separates machine learning from solving an optimization problem is that the former adopts a model using the training examples and then evaluates it on the test set to emulate the future performance on unseen data. This leads to the important distinction between the training error $MSE_{(train)}$

$$MSE_{train} = \frac{1}{m^{(train)}} \sum_{i=1}^{m^{(train)}} (y_i^{(train)} - \hat{y}_i^{(train)})^2 \tag{2.5}$$

and the test error $MSE_{(test)}$

$$MSE_{test} = \frac{1}{m^{(test)}} \sum_{i=1}^{m^{(test)}} (y_i^{(test)} - \hat{y}_i^{(test)})^2. \tag{2.6}$$

To gain a better understanding of the whole concept, a one-dimensional linear regression example is given in Fig. 2.2. The goal is to fit the model to the data points by reducing the distance between the regression line and the training examples. Then, as the algorithm is fed with unlabeled data points x_{new}, the fitted line is used for the prediction of y_{new}.

In the case of linear regression, it is possible to solve directly for the optimal model parameters w^* and b^* by setting the gradient of the cost function C to zero. This closed solution to the minimization problem is also known as normal equations [GBC16]. These are often expressed in a matrix form to simplify the handling of big

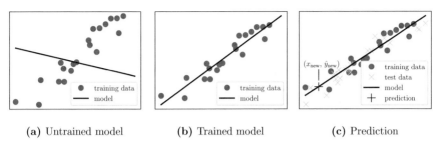

(a) Untrained model   (b) Trained model   (c) Prediction

Fig. 2.2 Linear regression example. Left: The initial model does not fit the training data. Middle: Shows the model properly fitted to the training data. Right: The trained model is used to make a prediction $\hat{y}_{new}$ for an unknown example x_{new}. Figures are inspired by [GBC16]

datasets. Then the problem can be written as

$$Xd = y, \tag{2.7}$$

where y is a vector with the y values of the dataset and $d = [w, b]^T$ consists of the weights and the bias that are to be determined. The matrix X contains the corresponding features evaluated at each x value. As the dataset is greater than the number of weights and the bias together, the system is overdetermined and the matrix X is thereby not symmetric and not invertible. Again, the goal is to minimize the residuum vector

$$r = y - Xd. \tag{2.8}$$

This is done by finding the sum of least squares and minimizing it by setting the derivative with respect to d to zero, which is discussed in further detail in [CK12]. This results in the following expression for the weights and the bias:

$$d = \begin{bmatrix} w \\ b \end{bmatrix} = (X^T X)^{-1} X^T y. \tag{2.9}$$

Unlike linear regression, a closed-form solution is not known for many practically relevant problems. In these cases, it is favorable to use a more generally applicable optimization technique. Section 2.6 introduces the gradient descent approach that lays the foundation for many optimization algorithms used in modern machine learning problems. Gradient descent techniques allow the iterative search for a minimum considering a large amount of parameters.

2.6 Optimization Techniques

Optimization plays an essential role in many machine learning algorithms. Generally, the goal of an optimization is to minimize or maximize an optimization criterion or objective function $C(\boldsymbol{\Theta})$ by altering $\boldsymbol{\Theta}$ [GBC16]. In the context of machine learning, this function is often called cost function, and $\boldsymbol{\Theta}$ denotes the parameters of a model, normally the weights $\boldsymbol{w}$ and biases b. A lot of research is dedicated to finding efficient methods that help to determine the optimal parameters $\boldsymbol{\Theta}^*$. When the optimization criterion is differentiable, a popular choice is the iterative gradient descent algorithm. A gradient descent approach finds a local minimum of a function by taking steps proportional to the negative gradient of the function at a given point. The gradient points in the direction of the steepest ascent, hence a small step in the opposite direction leads to a minimization of the function.

The algorithm is commonly used for neural networks but can be also applied to support vector machines or linear regression. In the latter case, the optimization criterion is convex meaning the function only has one global minimum. For neural networks, the optimization problem is non-convex and hence it is more likely to converge to a local minimum [GBC16].

As mentioned in Sect. 2.5, it is possible to analytically solve for the optimal parameters of a linear regression model. Nevertheless, it serves as a suitable example to explain the basic steps of gradient descent. To keep things simple, a linear regression model with a scalar input x, a corresponding weight w, and a scalar bias b is chosen

$$\hat{y} = wx + b. \tag{2.10}$$

Again, the goal is to find the optimal parameters w^* and b^* that minimize the mean squared error. The resulting cost function can be written as

$$C = \frac{1}{m} \sum_{i=1}^{m} (y_i - (wx_i + b))^2. \tag{2.11}$$

Now the partial derivatives for each parameter have to be calculated

$$\frac{\partial C}{\partial w} = \frac{1}{m} \sum_{i=1}^{m} -2x_i \left(y_i - (wx_i + b) \right),$$
$$\frac{\partial C}{\partial b} = \frac{1}{m} \sum_{i=1}^{m} -2 \left(y_i - (wx_i + b) \right). \tag{2.12}$$

If more parameters are involved, it is useful to rewrite the partial derivatives in the form of the gradient ∇C. The gradient is nothing else than a vector storing all the partial derivatives with respect to all parameters of a function. During each iteration step or epoch, to use machine learning terminology, the parameters of the model get updated according to the following rules:

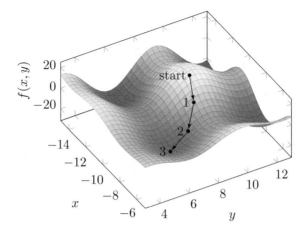

Fig. 2.3 Gradient descent on a surface defined by $f(x, y) = y \sin(x) + x \cos(y)$. Depending on the point of initialization gradient descent follows the steepest direction toward a minimum of the underlying function

$$w \leftarrow w - \alpha \frac{\partial C}{\partial w},$$
$$b \leftarrow b - \alpha \frac{\partial C}{\partial b}.$$

(2.13)

In the first step, the parameters are initialized as zero and then updated in each epoch until convergence is reached. The learning rate α controls the step size of each update [Bur19].

A graphical representation of gradient descent is depicted in Fig. 2.3. It illustrates how the algorithm follows the direction of steepest descent toward a minimum. The outcome is highly dependent on the initialization of the starting point. If the starting point is initialized at a different location, the algorithm might descend into another local minimum. This shows the difficulties to find a global minimum in non-convex optimization problems.

A major drawback of gradient descent is its sensitivity to the choice of the learning rate α. If α is too large, the algorithm starts to oscillate or even fails to converge at all. Conversely, a small α value leads to an extremely low convergence rate [Bur19].

Another issue is that with an increasing number of examples, the computational costs for each epoch grow as well because all partial derivatives are evaluated for the whole training set of size m. Since the whole training set is used, the method described above is often referred to as full-batch gradient descent. Alternatively, the parameter update can be performed instantly after computing the gradient for a single example in the training data. This introduces stochasticity to the algorithm because the gradient of a single example might indicate a substantially different direction than the gradient computed for the whole training batch [GBC16]. However, in practice, stochastic or sometimes called online gradient descent often shows better convergence properties, especially when combined with an adaptive learning rate. A popular and widely used compromise between full-batch and stochastic gradient descent (SGD) is mini-batch gradient descent. The idea is to find an approximation of the gradient by evaluating the gradient just for a small sample of the data, a so-called mini-batch. Instead of evaluating the gradient for the whole training set, it is replaced by an estimator that was computed on a sample with a fixed size. The differences between the implementation of the three possible optimization strategies are described in Sect. 3.6 in the context of neural networks.

Other improved versions of stochastic gradient descent include AdaGrad, which automatically adapts the learning rate, and the momentum method that accelerates stochastic gradient descent by selecting the relevant direction and thus reduces oscillations [Bur19].

2.7 Overfitting Versus Underfitting

The previous section has introduced the test error, in particular, the MSE_{test}, to measure how well a model is expected to perform when confronted with new inputs. The ability to generate good predictions for previously unseen data is called generalization and the associated error is called the generalization error. Therefore, the test error is also considered to be an estimate of the generalization error.

Supervised learning algorithms like linear regression are based on the idea that a low training error also leads to a small test error. Statistical learning theory provides some assumptions to support this idea [GBC16]:

1. The data contains all necessary information to solve the problem.
2. Examples in each dataset are independent of each other.
3. The training and test sets are identically distributed, meaning each example is generated with the same probability distribution.

Assuming all given statements are true, the training error ought to be the same as the test error. Since in reality, the data generating probability distribution is unknown a priori, the test error is usually higher than the training error. Figure 2.4 illustrates the two main reasons for that. Firstly, the capacity of a model can be too high, meaning the algorithm chooses a complex function that fits the training data perfectly, but fails to generalize, because it overestimates the importance of noisy data. This behavior is called overfitting (cf. Fig. 2.4c). Secondly, a model with a very low capacity is applied to a highly non-linear problem. For example, if the hypothesis space of a model only contains linear functions, it is not able to represent data that is following quadratic or cubic functions. This case describes underfitting (cf. Fig. 2.4a).

The capacity of a model, more precisely, its ability to fit a wide variety of functions, plays an important role in terms of performance. Choosing the capacity in a way that suits the complexity of the problem or, in other words, finding the balance between underfitting and overfitting is a central challenge in machine learning [GBC16]. The relationship between the training error and the generalization error with respect to

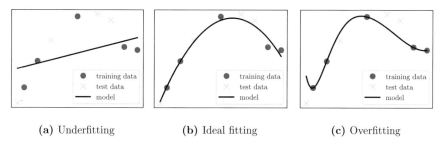

(a) Underfitting   (b) Ideal fitting   (c) Overfitting

Fig. 2.4 Representation of the underfitting and overfitting problem on the example of a one-dimensional regression. Figures inspired by Goodfellow [GBC16] and Burkov [Bur19]

the capacity is depicted in Fig. 2.5. As indicated, the optimal capacity is reached when the generalization error is as low as possible.

A way to tackle the problem of underfitting (cf. Fig. 2.4a) is to increase the set of functions an algorithm is allowed to select. Sticking with the example of linear regression, assuming a simple model with a single input x we have

$$\hat{y} = wx + b \tag{2.14}$$

as in Eq. (2.10). The model can be easily extended to include polynomials. Adding x^2 as a new feature, a second-order polynomial regression of the following form is obtained:

$$\hat{y} = w_1 x^2 + w_2 x + b, \tag{2.15}$$

where $x_1 = x^2$ and $x_2 = x$. Introducing more polynomials to the model not only increases the number of features x_i, but also introduces the same amount of parameters w_i. Hence, the algorithm has more possibilities to tune and adapt the model to fit the target function appropriately (cf. Fig. 2.4b). Yet, if the polynomial degree and therefore the capacity becomes too large, the model starts to overfit [GBC16]. This is shown in Fig. 2.4c for the case of regression with a polynomial degree of seven

$$\hat{y} = \sum_{i=1}^{7} \left(w_i x^i \right) + b. \tag{2.16}$$

In the literature, over- and underfitting are often associated with the terms bias and variance. A high bias means the model produces many mistakes on the training set, so it is equivalent to underfitting. Coming from statistics, the term variance describes the model's sensitivity to changes in the dataset. If the variance is high, small deviations in the training data result in a very different model. This behavior is closely related to overfitting [Bur19].

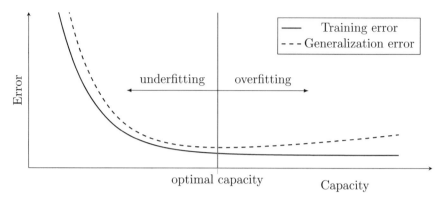

Fig. 2.5 Generalization error and training error in relation with the capacity of a model. Figure inspired by Fortman-Roe, Goodfellow, and Mehta et al. and see [For12, GBC16, Meh+19]

2.8   Regularization

Overcoming underfitting or overfitting, i.e. finding the appropriate capacity for a model, is one of the most challenging tasks in machine learning. The previous section showed a possibility to deal with underfitting by increasing the number of features x_i. In the same way, overfitting can be reduced when features are removed from the model. However, this option is not very popular, since removing features means that information about the problem gets lost. Another countermeasure against overfitting is adding more examples to the training set [GBC16]. In most cases, it is not feasible or simply impossible to gather more data about the problem. Conversely, removing training data can help in the case of underfitting. Yet again, this leads to a loss of information. The influence of the training set size on the error measures is illustrated in Fig. 2.6.

Since the complexity of the underlying problem is usually unknown, it is hardly possible to choose a model with the appropriate capacity beforehand. So another approach to overcome overfitting is to retain a high capacity, but to introduce a control mechanism called regularizer that prefers the selection of certain functions over others. For instance, the cost function for linear regression (cf. Eq. (2.3)) can be modified to include an additional term that penalizes large weights

$$\tilde{C}(\boldsymbol{w}, b) = C(\boldsymbol{w}, b) + \lambda ||\boldsymbol{w}||_1, \tag{2.17}$$

where $||\boldsymbol{w}||_1$ denotes the L^1-norm of the weight vector $\boldsymbol{w}$. Adding a term proportional to the magnitude of the weights forces the algorithm to select a model with smaller weights. In case of linear or polynomial regression, the weights can be understood as the coefficients determining the slope of the function. For instance, the model depicted in Fig. 2.7c exhibits large oscillations, which indicates an underlying function with large slope coefficients. Hence, penalizing large weights is a way to reduce

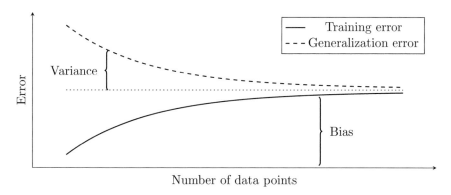

Fig. 2.6 Influence of dataset size on training and generalization error. The larger the amount of available data, the closer both error measurements get. The distance between the training error and the asymptote marked in red can be interpreted as the bias. In the same way, the distance between die generalization error and the asymptote denotes the variance. Figure inspired by Abu-Mostafa et al. [AML12] and Mehta et al. [Meh+19]

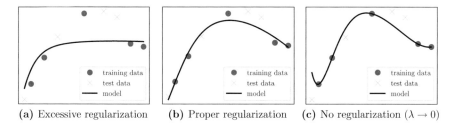

(a) Excessive regularization   **(b)** Proper regularization   **(c)** No regularization ($\lambda \to 0$)

Fig. 2.7 Regularization. Figures inspired by Goodfellow [GBC16] and Bishop [Bis06]

the slope coefficients and the occurrence of oscillations. The influence of the penalty term is controlled by λ. If λ equals zero, the algorithm simply returns the initial linear regression model (cf. Fig. 2.7c). Contrarily, a large value for λ leads to extremely small weights, which basically eliminates the corresponding features resulting in a very simple or sparse model (cf. Fig. 2.7a). In practice, this technique called L^1 or lasso regularization is used for feature selection, a method to remove unimportant features from a model. A similar approach is L^2 or ridge regularization which adds the square of the L^2-norm of the weights instead of the sum of the absolute values. For linear regression, the regularized cost function can be written as

$$\tilde{C}(\boldsymbol{w}, b) = C(\boldsymbol{w}, b) + \lambda \boldsymbol{w}^T \boldsymbol{w}. \tag{2.18}$$

An advantage of L^2 regularization is that the penalty term is differentiable [Bur19]. This gains more importance when used in combination with the gradient-based optimization algorithms, which are introduced in Sect. 2.6.

Apart from model selection and adding a penalty term to the objective function, there exist many other techniques to implicitly and explicitly express preferences for a certain solution. A few more are explained in Chap. 3 about neural networks. Overall, these approaches can be summarized under the term regularization. A definition by Goodfellow states: "Regularization is any modification we make to a learning algorithm that is intended to reduce its generalization error (low variance) but not its training error (low bias) [GBC16]." Accordingly, regularization is also known as the bias-variance trade-off [Bur19].

When applying L^1 or L^2 regularization to a model, a new parameter λ is added to the objective function. In contrast to weights and bias, λ is not part of the optimization objective and thus has to be set manually by the user. Generally, all external settings controlling the behavior of an algorithm are called hyperparameters [Bur19]. They give the opportunity to further improve the performance by doing a so-called hyperparameter tuning. For this purpose, a new subset from the training data has to be selected, since the test set cannot be used for any parameter optimization. Therefore, the training data is split into a set for training the standard model parameters and a validation set used for tuning the hyperparameters.

References

[Bur19]   Andriy Burkov. *The Hundred-Page Machine Learning Book*. Andriy Burkov, Jan. 13, 2019. 160 pp. ISBN: 978-1-9995795-0-0.

[Mit97]   Tom M. Mitchell. *Machine Learning*. McGraw-Hill series in computer science. New York: McGraw-Hill, 1997. 414 pp. ISBN: 978-0-07-042807-2.

[GBC16]   Ian Goodfellow, Yoshua Bengio, and Aaron Courville. *Deep Learning*. MIT Press, 2016. ISBN: 0-262-03561-8. URL: http://www.deeplearningbook.org.

[Ng20]   Andrew Ng. "Machine Learning". Online course. Online course. 2020. URL: https://www.coursera.org/learn/machine-learning?.

[Cho18]   François Chollet. *Deep learning with Python*. OCLC: ocn982650571. Shelter Island, New York: Manning Publications Co, 2018. 361 pp. ISBN: 978-1-61729-443-3.

[Sil+17]   David Silver et al. "Mastering the game of Go without human knowledge". In: *Nature* 550.7676 (Oct. 2017), pp. 354–359. ISSN: 0028-0836, 1476–4687. https://doi.org/10.1038/nature24270. URL: http://www.nature.com/articles/nature24270 (visited on 03/16/2020).

[CK12]   Ward E. Cheney and David R. Kincaid. *Numerical Mathematics and Computing*. 7th. Cengage Learning, 2012. 704 pp. ISBN: 978-1-133-10371-4.

[For12]   Scott Fortmann-Roe. *Understanding the Bias-Variance Tradeoff*. http://scott.fortmannroe.com/docs/BiasVariance.html. 2012.

[Meh+19]   Pankaj Mehta et al. "A high-bias, low-variance introduction to Machine Learning for physicists". In: *Physics Reports* 810 (May 2019), pp. 1–124. ISSN: 03701573. https://doi.org/10.1016/j.physrep.2019.03.001. arXiv:1803.08823. URL: http://arxiv.org/abs/1803.08823 (visited on 01/14/2020).

[AML12]   Yaser S. Abu-Mostafa, Malik Magdon-Ismail, and Hsuan-Tien Lin. *Learning From Data*. S.l.: AMLBook, 2012. 213 pp. ISBN: 978-1-60049-006-4.

[Bis06]   Christopher M. Bishop. *Pattern recognition and machine learning*. Information science and statistics. New York: Springer, 2006. 738 pp. ISBN:978-0-387-31073-2.

Chapter 3
Neural Networks

For supervised learning tasks, artificial neural networks (ANNs) are the state-of-the-art algorithmic architecture [Mar19]. Inspired by the neurons of the brain, an early form, known as perceptron, was created by Frank Rosenblatt in 1958. After the late 1960s, the development stagnated due to the lack of computational power and efficient methods for network training. With the introduction of the backpropagation algorithm in 1986, the learning capabilities of neural networks improved significantly. Nevertheless, their application only became practical in the early 2000s on the hardware available at this time. A major breakthrough leading to more attention was the success of a deep neural network by Krizhevsky et al. that won the image recognition challenge ImageNet in 2012 by a large margin [Dem+14]. The success was also driven by developments in computer graphics that allowed the exploitation of graphical processing units (GPUs) for the training process.

Since then, artificial neural networks have been applied to solve a great variety of problems. Typical tasks include speech, image, and natural language processing; autonomous driving; playing board and computer games; algorithmic trading; or weather forecasting. Recently, also physicists and engineers started to investigate how conventional methods can benefit from the capabilities of neural networks [Car+19]. Even though fundamentals in this still-young area of research have been established, most results of modern neural networks are based on empirical studies and heuristics [Meh+19]. This chapter introduces the simplest form of neural networks and explains its fundamentals in an illustrative example. Sections 3.10.1 and 3.10.2 are devoted to more advanced neural network architectures that were designed to perform exceptionally well in specific tasks such as image or speech recognition, language translation, or time series forecasting.

Electronic supplementary material The online version of this chapter (https://doi.org/10.1007/978-3-030-76587-3_3) contains supplementary material, which is available to authorized users.

S. Kollmannsberger et al., *Deep Learning in Computational Mechanics*,
Studies in Computational Intelligence 977,
https://doi.org/10.1007/978-3-030-76587-3_3

3.1   Feed-Forward Neural Network

When seen as a black-box, a neural network is like any other supervised learning model just a parameterized function defining a mapping $y = f_{NN}(x)$. A par-
ticularity of neural networks is that they are typically composed of many nested
functions. For instance, three functions f_1, f_2, and f_3 might form the mapping
$f_{NN}(x) = f_3(f_2(f_1(x)))$, where each function f_l represents a layer of the network.
The information in form of input x flows from the input layer through an arbitrary
number of so-called "hidden" layers to the output layer, hence the name feed-forward
neural network. The number of layers defines the depth, whereas the amount of hidden neurons determines the width of a network. In this context, the term "deep"
learning often refers to neural networks with more than one hidden layer [GBC16].
A simple network architecture is depicted in Fig. 3.1. The circles represent the basic
units called "neurons", whereas the connections can be interpreted as weights that
control the importance of their inputs. If every neuron from the previous layer is connected with each neuron of the next layer, then the neural network is said to be "fully
connected". A single neuron receives a vector of features x and produces a scalar
output $a(x)$, which serves as an input for the neurons in the next layer [Meh+19]. The
output of a neuron can be decomposed into two operations. First, the input vector x
is transformed into

$$z = w^T x + b \tag{3.1}$$

with a neuron-specific weight w and bias b. Second, the neuron's output $a(x)$ is
computed by applying the non-linear function σ to the resulting scalar z (cf. Fig. 3.2)

$$a(x) = \sigma(z). \tag{3.2}$$

The function σ is called activation function and is usually chosen to be the same for
all neurons [GBC16]. Without the use of a non-linear activation function, the model

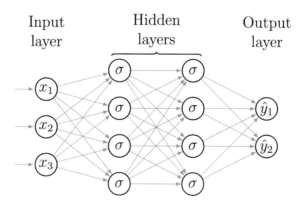

Fig. 3.1 A fully connected feed-forward neural network

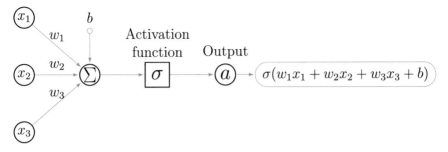

Fig. 3.2 A single neuron with three inputs

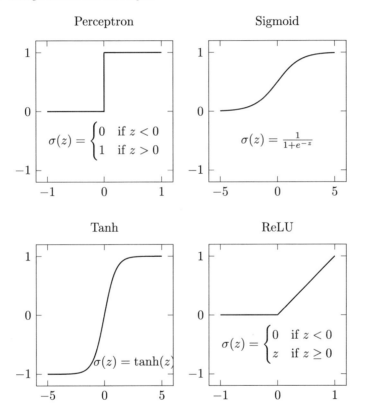

Fig. 3.3 Common activation functions

would not be able to represent any non-linearities in the data. The reason is that, regardless of how many linear transformations are applied to an input, the output would still remain a linear function of the input. Typical candidates for the activation $\sigma(z)$ are the hyperbolic tangent, the sigmoid function, and the rectified linear unit, in short ReLU (see Fig. 3.3). A more detailed discussion about the properties of the different activation functions is provided in Sect. 3.5.

Depending on the function chosen for the output neurons, neural networks are able to solve either regression or classification tasks. For regression, simply applying a linear function in the output layer leads to a result in form of a real number. In the case of classification, a common choice is the sigmoid function to enforce a binary encoding of the output that can be easily transformed into labels.

Generally, a fully connected feed-forward neural network with one hidden layer is capable of approximating any continuous multi-input/multi-output function with arbitrary precision, given that the hidden layer contains a sufficient amount of neurons [Meh+19]. Known as the universal approximation theorem, this hypothesis was formally proven by multiple researchers independently, e.g. in "Approximation by superpositions of a sigmoidal function" by Cybenko, just to mention one of the first references [Cyb89]. A graphical and very intuitive explanation of the universality theorem can be found in chapter four of Nielsen's online book [Nie15, Chap. 4]. The principal idea is that hidden neurons allow the generation of step functions with arbitrary heights and offsets that can be superpositioned to construct any continuous function [Meh+19]. However, the whole concept is rather of theoretical importance. In practice, the usage of deep networks with multiple hidden layers is preferred. Multi-layer architectures exhibit the same representational power as comparable wide networks with a single hidden layer, while being computationally more efficient in the training process, as a paper by Mhaskar et al. suggested [MLP16]. Nevertheless, this topic is controversially discussed and continues to be an area of active research [Meh+19].

When dealing with discontinuous functions, a feed-forward neural network can provide a sufficient continuous approximation [Nie15]. Nevertheless, their applicability for discontinuous problems is limited. The solution often exhibits oscillations around the discontinuities similar to the Gibbs phenomenon observed in Fourier series approximations of discontinuous functions [HH79]. In "Constructive Approximation of Discontinuous Functions by Neural Networks", Llanas et al. propose a way to overcome this shortcoming and show an almost uniform approximation of a piece-wise continuous function by a single hidden-layer feed-forward neural network [LLS08].

In order to give an illustrative example of the representational capabilities, a neural network is used to generate color plots. The outputs for different numbers of hidden layers are compared in Fig. 3.4. All networks have two inputs, represented by the horizontal and vertical image axis as well as the corresponding output that controls the color value of the image. With an increasing number of layers, the network is able to show more complex functions. These images were generated with random weights and biases and the networks have not been trained for approximating a specific function or image yet.

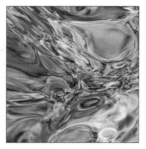

(a) 1 hidden layer with 100 neurons

(b) 10 hidden layers with 100 neurons each

(c) 20 hidden layers with 100 neurons each

Fig. 3.4 Output visualization of a multi-layer neural network. It has two inputs corresponding to the x and y coordinates of a picture and one output defining the color at the pixel (x, y). The parameters of the network are randomly initialized and are not trained. The code used for generating the pictures was developed by Marquardt [Mar17]

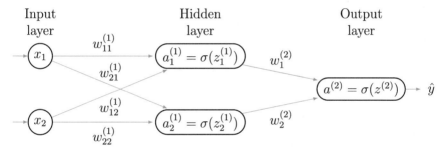

Fig. 3.5 A simple feed-forward neural network example

3.2 Forward Propagation

Before proceeding with the training process of a neural network, it is important to get a better understanding of how the data is propagated from the input to the output layers. A very simple example network with two inputs x_1 and x_2, one hidden layer, and a single output $\hat{y}$ is depicted in Fig. 3.5.

The goal is to compute the prediction $\hat{y}$ of the neural network for one given example

$$x = \begin{bmatrix} x_1 \\ x_2 \end{bmatrix} = \begin{bmatrix} 3 \\ 2 \end{bmatrix}$$

and weights

$$W^{(1)} = \begin{bmatrix} w_{11}^{(1)} & w_{12}^{(1)} \\ w_{21}^{(1)} & w_{22}^{(1)} \end{bmatrix} = \begin{bmatrix} 1 & -3 \\ -2 & 1 \end{bmatrix}, \quad w^{(2)} = \begin{bmatrix} w_1^{(2)} \\ w_2^{(2)} \end{bmatrix} = \begin{bmatrix} 2 \\ -1 \end{bmatrix}.$$

For the sake of simplicity, the biases for each neuron are set to zero and thus are neglected in the following calculations. Furthermore, the activation function for all neurons is chosen to be $\sigma(z) = (z_i)^2$. The input vector x is fed to the network and every entry is stored in the corresponding neuron of the input layer without any further modifications. As a first step toward computing the output of the hidden layer, the weighted sum of the inputs is calculated and stored in the variable $z^{(1)}$

$$z^{(1)} = W^{(1)}x = \begin{bmatrix} 1 & -3 \\ -2 & 1 \end{bmatrix} \begin{bmatrix} 3 \\ 2 \end{bmatrix} = \begin{bmatrix} -3 \\ -4 \end{bmatrix}.$$

Then, the activation function σ is applied element-wise to the resulting vector $z^{(1)}$ to get the output $a^{(1)}$ of the hidden layer

$$a^{(1)} = \sigma(z^{(1)}) = \begin{bmatrix} \sigma(-3) \\ \sigma(-4) \end{bmatrix} = \begin{bmatrix} 9 \\ 16 \end{bmatrix}.$$

Now, the output $a^{(1)}$ serves as the input for the neurons of the next layer. Again, the linear transformation $z^{(1)}$ is computed first

$$z_1^{(2)} = (w^{(2)})^T a^{(1)} = \begin{bmatrix} 2 & -1 \end{bmatrix} \begin{bmatrix} 9 \\ 16 \end{bmatrix} = 2.$$

Finally, the prediction of the neural network $\hat{y}$ is computed by applying the activation function to $z_1^{(2)}$

$$a_1^{(2)} = \sigma(z_1^{(2)}) = \sigma(2) = 4 = \hat{y}.$$

To generalize the computations from the example, the output for the current neuron j in the lth layer is calculated by the following formula [Nie15]:

$$a_j^{(l)} = \sigma(z_j^{(l)}) = \sigma\left(\sum_k w_{jk}^{(l)} a_k^{(l-1)} + b_j^{(l)}\right), \tag{3.3}$$

where $a_k^{(l-1)}$ is the output of the kth neuron from the previous layer $(l-1)$ and $w_{jk}^{(l)}$ is the weight connecting the kth neuron with the current neuron j. After adding the neuron-specific bias $b_j^{(l)}$ to the weighted sum of all outputs k from the previous layer, the non-linear activation function is applied [Nie15].

Using only matrix and vector notation, Eq. (3.3) yields [Nie15]

$$a^{(l)} = \sigma(z^{(l)}) = \sigma(W^{(l)} a^{(l-1)} + b^{(l)}). \tag{3.4}$$

The images in Fig. 3.4 were generated in a very similar fashion, namely by forward-propagating arbitrary inputs through a feed-forward neural network with random weights and biases. More complex images were achieved only by adding more hidden layers and neurons to the network architecture.

3.3 Differentiation

Up to this point, the neural network was only able to transform an input into an output depending on the randomly initialized set of weights and biases. Meaning the network has not utilized any data to learn the optimal parameters w^* and b^* that generate the desired output. As for the example of linear regression, learning from data can be formulated as an optimization problem that requires the definition of a suitable cost function. Once again the mean squared error loss is chosen to quantify the prediction accuracy of the model. The cost function takes on the familiar form

$$C = \frac{1}{2m} \sum_{i=1}^{m} (y_i - \hat{y}_i)^2, \tag{3.5}$$

but introducing the factor $\frac{1}{2}$ to simplify the expression of the derivative needed for later calculations. To find optimal network parameters, the cost function is minimized with the gradient descent method that was introduced in Sect. 2.6. Essential for the iterative parameter update (cf. Eq. (2.13)) is the computation of the gradient of the cost function with respect to the weights and biases.

For simple network architectures, it is possible to compute the corresponding derivatives analytically. Therefore, the network must be expressed as a function of the inputs and the network parameters. Given the example from the previous section (cf. Fig. 3.5), the analytic expression of the network is

$$\hat{y} = \sigma \left(w_1^{(2)} \sigma(w_{11}^{(1)} x_1 + w_{12}^{(1)} x_2) + w_2^{(2)} \sigma(w_{21}^{(1)} x_1 + w_{22}^{(1)} x_2) \right). \tag{3.6}$$

To improve the readability and since the analytical form is valid for all examples, the index i is dropped here. Now the cost function for a single example of the network from Fig. 3.5 can be written as

$$C = \frac{1}{2}(y - \hat{y})^2 = \frac{1}{2} \left(y - \sigma(w_1^{(2)} \sigma(w_{11}^{(1)} x_1 + w_{12}^{(1)} x_2) + w_2^{(2)} \sigma(w_{21}^{(1)} x_1 + w_{22}^{(1)} x_2)) \right)^2. \tag{3.7}$$

To compute the analytical expression, e.g. for the derivative $\frac{\partial C}{\partial w_{11}^{(1)}}$, the chain rule is applied to Eq. (3.5) so that

$$\frac{\partial C}{\partial w_{11}^{(1)}} = \frac{\partial C}{\partial \hat{y}} \frac{\partial \hat{y}}{\partial w_{11}^{(1)}}, \tag{3.8}$$

with

$$\frac{\partial C}{\partial \hat{y}} = -(y - \hat{y}). \tag{3.9}$$

Deriving Eq. (3.6) with respect to $w_{11}^{(1)}$ yields

$$\frac{\partial \hat{y}}{\partial w_{11}^{(1)}} = \sigma' \left(w_1^{(2)} \sigma(w_{11}^{(1)} x_1 + w_{12}^{(1)} x_2) + w_2^{(2)} \sigma(w_{21}^{(1)} x_1 + w_{22}^{(1)} x_2) \right) w_1^{(2)} \sigma'(w_{11}^{(1)} x_1 + w_{12}^{(1)} x_2) x_1.$$

(3.10)

Similarly, the remaining derivatives could be computed, but for larger and deeper networks, the analytical differentiation becomes infeasible. In engineering applications, a common approach for the numerical approximation of derivatives is the method of finite differences. However, this approach introduces truncation errors and quickly becomes computationally expensive for large networks with many parameters [Bay+18]. Automatic differentiation is a more efficient alternative allowing the fast and precise numerical evaluation of higher-order derivatives. Generally, a computer program is nothing else than the sum of its basic arithmetic operations and elementary function evaluations. Automatic differentiation augments the code during execution by immediately evaluating and storing the derivatives of each basic operation. Then, repeatedly applying the chain rule to the accumulated derivative values allows to compute the derivative of the whole composition [Bay+18].

3.4   Backpropagation

Backpropagation is an efficient technique for determining the partial derivatives of graph-structured functions and in particular neural networks. It belongs to the field of automatic differentiation that deals with the algorithmic computation of derivatives [GBC16]. The idea of backpropagation was reinvented many times and is better known as reverse-mode differentiation in other scientific fields [Bay+18]. In the following, the general form of the backpropagation algorithm is derived and then illustrated on the example network from Sect. 3.2 (cf. Fig. 3.5).

If the cost function C is seen as an average over the sum of cost functions C_i of the individual examples i

$$C = \frac{1}{m} \sum_{i=1}^{m} C_i,$$

(3.11)

then the cost C_i for one example i can be written as

$$C_i = \frac{1}{2}(y_i - \hat{y}_i)^2 = \frac{1}{2}(y_i - a_i^{(L)})^2,$$

(3.12)

where $a_i^{(L)}$ denotes the output of the last layer L and simultaneously the output $\hat{y}_i$ of the neural network. From here, the computation of the derivatives is only performed for a single example. Thus, the index i in Eq. (3.12) is dropped to increase the readability. As mentioned, the goal of the backpropagation algorithm is to compute the partial derivatives of the cost function with respect to the weights $\frac{\partial C}{\partial w_{jk}^{(l)}}$ and biases $\frac{\partial C}{\partial b_j^{(l)}}$.

Recalling Eq. (3.3), the output of a layer l is defined as

$$a_j^{(l)} = \sigma(z_j^{(l)}), \tag{3.13}$$

where

$$z_j^{(l)} = \sum_k w_{jk}^{(l)} a_k^{(l-1)} + b_j^{(l)} \tag{3.14}$$

is computed using the outputs of the previous layer $a_k^{(l-1)}$.

Now, rewriting $\frac{\partial C}{\partial w_{jk}^{(l)}}$ yields

$$\frac{\partial C}{\partial w_{jk}^{(l)}} = \frac{\partial C}{\partial z_j^{(l)}} \frac{\partial z_j^{(l)}}{\partial w_{jk}^{(l)}} = \delta_j^{(l)} \frac{\partial z_j^{(l)}}{\partial w_{jk}^{(l)}} = \delta_j^{(l)} a_k^{(l-1)}, \tag{3.15}$$

with

$$\frac{\partial z_j^{(l)}}{\partial w_{jk}^{(l)}} = \frac{\partial}{\partial w_{jk}^{(l)}} \sum_h w_{jh}^{(l)} a_h^{(l-1)} + b_j^{(l)} = a_k^{(l-1)}, \tag{3.16}$$

where the index h runs over all neurons from the previous layer $(l-1)$. Similarly, $\frac{\partial C}{\partial b_j^{(l)}}$ is computed as

$$\frac{\partial C}{\partial b_j^{(l)}} = \frac{\partial C}{\partial z_j^{(l)}} \frac{\partial z_j^{(l)}}{\partial b_j^{(l)}} = \delta_j^{(l)} \frac{\partial z_j^{(l)}}{\partial b_j^{(l)}} = \delta_j^{(l)}, \tag{3.17}$$

since

$$\frac{\partial z_j^{(l)}}{\partial b_j^{(l)}} = \frac{\partial}{\partial b_j^{(l)}} \sum_k w_{jk}^{(l)} a_k^{(l-1)} + b_j^{(l)} = 1. \tag{3.18}$$

Equations (3.15) and (3.17) introduce a new variable $\delta_j^{(l)}$ that describes the sensitivity of the cost function toward a change in the neuron's weighted input $z_j^{(l)}$. If $\delta_j^{(l)}$ for each neuron j in layer l is known, the derivative $\frac{\partial C}{\partial w_{jk}^{(l)}}$ can be computed simply by multiplying $\delta_j^{(l)}$ of the jth neuron of the current layer l with the output $a_k^{(l-1)}$ of the kth neuron from the previous layer $(l-1)$. The derivative with respect to the bias $\frac{\partial C}{\partial b_j^{(l)}}$ directly equates to $\delta_j^{(l)}$.

In the next step, an expression for $\delta_j^{(L)}$ in the output layer L is derived

$$\delta_j^{(L)} = \frac{\partial C}{\partial z_j^{(L)}} = \frac{\partial C}{\partial a_j^{(L)}} \frac{\partial a_j^{(L)}}{\partial z_j^{(L)}} = \frac{\partial C}{\partial a_j^{(L)}} \sigma'(z_j^{(L)}) = -(y_j - \sigma(z_j^{(L)}))\sigma'(z_j^{(L)}), \tag{3.19}$$

recalling that $a_j^{(L)} = \sigma(z_j^{(L)})$ and inserting

$$\frac{\partial a_j^{(L)}}{\partial z_j^{(L)}} = \frac{\partial}{\partial z_j^{(L)}} \sigma(z_j^{(L)}) = \sigma'(z_j^{(L)}). \tag{3.20}$$

What is still missing is a general term for $\delta_j^{(l)}$ in an arbitrary layer l. Utilizing the chain rule $\delta_j^{(l)}$ can be expressed in terms of $\delta_k^{(l+1)}$

$$\delta_j^{(l)} = \frac{\partial C}{\partial z_j^{(l)}} = \sum_k \frac{\partial C}{\partial z_k^{(l+1)}} \frac{\partial z_k^{(l+1)}}{\partial z_j^{(l)}} = \sum_k \frac{\partial z_k^{(l+1)}}{\partial z_j^{(l)}} \delta_k^{(l+1)}. \tag{3.21}$$

With $a_j^{(l)} = \sigma(z_j^{(l)})$ and the definition of $z_k^{(l+1)}$ being

$$z_k^{(l+1)} = \sum_j w_{kj}^{(l+1)} a_j^{(l)} + b_k^{(l+1)} = \sum_j w_{kj}^{(l+1)} \sigma(z_j^{(l)}) + b_k^{(l+1)}, \tag{3.22}$$

the derivative with respect to $z_j^{(l)}$ can be calculated as

$$\frac{\partial z_k^{(l+1)}}{\partial z_j^{(l)}} = w_{kj}^{(l+1)} \sigma'(z_j^{(l)}). \tag{3.23}$$

Substituting Eq. (3.23) into Eq. (3.21) finally yields

$$\delta_j^{(l)} = \sum_k w_{kj}^{(l+1)} \delta_k^{(l+1)} \sigma'(z_j^{(l)}). \tag{3.24}$$

With Eqs. (3.15), (3.19), and (3.24), all expressions to calculate the partial derivatives of the simple example network are available. Since the biases in the example are chosen to be zero, the computation of the derivatives has to be executed with respect to the four entries of $W^{(1)}$ and two entries of $W^{(2)}$. This also implies that Eq. (3.17) is not needed. The single data point considered in this example consists of the input vector $x = \begin{bmatrix} 3 \\ 2 \end{bmatrix}$ and the target $y = 1$. In order to compute the delta for the output neuron $\delta^{(L)}$, the derivative of the activation function $\sigma'(z) = 2z$ and the results for $z_1^{(2)}$ from the previous section are inserted into Eq. (3.19) resulting in

$$\delta_1^{(2)} = -(y - \sigma(z_1^{(2)}))\sigma'(z_1^{(2)}) = -(1 - 4)\sigma'(2) = 12.$$

This intermediate result is then propagated backward to compute $\delta_1^{(1)}$ and $\delta_2^{(1)}$ of the hidden layer using Eq. (3.24)

$$\delta_1^{(1)} = w_1^{(2)} \delta_1^{(2)} \sigma'(z_1^{(1)}) = 2 * 12 * \sigma'(-3) = -144,$$

$$\delta_2^{(1)} = w_2^{(2)} \delta_1^{(2)} \sigma'(z_2^{(1)}) = (-1) * 12 * \sigma'(-4) = 96.$$

Finally, the partial derivatives are calculated with Eq. (3.15) using the previously computed δ-values

$$\frac{\partial C}{\partial w_{11}^{(1)}} = \delta_1^{(1)} x_1 = -144 * 3 = -432,$$

$$\frac{\partial C}{\partial w_{12}^{(1)}} = \delta_1^{(1)} x_2 = -144 * 2 = -288,$$

$$\frac{\partial C}{\partial w_{21}^{(1)}} = \delta_2^{(1)} x_1 = 96 * 3 = 288,$$

$$\frac{\partial C}{\partial w_{22}^{(1)}} = \delta_2^{(1)} x_2 = 96 * 2 = 192,$$

$$\frac{\partial C}{\partial w_1^{(2)}} = \delta_1^{(2)} a_1^{(1)} = 12 * 9 = 108,$$

$$\frac{\partial C}{\partial w_2^{(2)}} = \delta_1^{(2)} a_2^{(1)} = 12 * 16 = 192.$$

The partial derivatives can now be used to update the parameters according to the update rule from Eq. (2.13) in Sect. 2.6:

$$W_{new}^{(1)} = W^{(1)} - \alpha \nabla C_{W^{(1)}}, \tag{3.25}$$
$$w_{new}^{(2)} = w^{(2)} - \alpha \nabla C_{w^{(2)}}. \tag{3.26}$$

Updating the weights with a learning rate $\alpha = 0.001$ yields

$$W_{new}^{(1)} = \begin{bmatrix} 1 & -3 \\ -2 & 1 \end{bmatrix} - 0.001 \begin{bmatrix} -432 & -288 \\ 288 & 192 \end{bmatrix} = \begin{bmatrix} 1.432 & -2.172 \\ -2.288 & 0.808 \end{bmatrix} \tag{3.27}$$

and

$$w_{new}^{(2)} = \begin{bmatrix} 2 \\ -1 \end{bmatrix} - 0.001 \begin{bmatrix} 108 \\ 192 \end{bmatrix} = \begin{bmatrix} 1.892 \\ -1.192 \end{bmatrix}. \tag{3.28}$$

An intuitive explanation of backpropagation can be found in the second chapter of Nielsen's online book "Neural Networks and Deep Learning" that was used as a reference for the formal derivation of the presented algorithm [Nie15, Chap. 5]. Another illustrative example is described in Christopher Olah's blog post [Ola15a].

3.5   Activation Function

As observed in the preceding section, the activation function needs to be differentiable in order to train neural networks with gradient-based methods. Inspired by the biological archetype, the first neural networks, known as perceptrons, used a Heaviside step function to imitate an active ($= 1$) or inactive neuron ($= 0$) (see Fig. 3.3). However, the derivative of a step function is zero everywhere except at the jump and thus makes it impractical for training a network. Until recently, continuous functions like the hyperbolic tangent or the sigmoid, also known as the logistic function, were popular choices for the activation σ. Unfortunately, with increasing weights, those functions tend to saturate, resulting in very small gradients that prevent the network from further learning [GBC16]. To resolve this problem of "vanishing gradients", a new type of activation function was introduced. Rectified linear units or in short ReLU (cf. Fig. 3.3) are now the default recommendation for the use with most feed-forward neural networks [GBC16]. They help to reduce saturation, since the function grows linearly for positive inputs. Generally, the choice of the activation function has a considerable impact on the performance and especially on the learning rate of a model, but usually depends on the specific problem and the experience of the machine learning practitioner [Ng20].

3.6   Learning Algorithm

Within the previous sections, all relevant information has been provided to build a neural network architecture. To summarize, the following aspects are required to train the network for a supervised learning task [Cho18]:

- input data X and corresponding targets y, divided into a test, training, and validation set;
- network topology, defined by the input, output, and hidden layers, the corresponding number of neurons, and their connections;
- an activation function σ;
- a loss function C, which defines the feedback signal used for learning;
- a way to compute the gradients w.r.t the network parameters, e.g. backpropagation;
- an optimizer with learning rate α, which determines how learning proceeds.

Algorithm 1 roughly describes the learning algorithm for a feed-forward neural network that regresses a scalar value $\hat{y}$ from a given input x.

Instead of using full-batch gradient descent, the parameters can be updated with a stochastic or mini-batch gradient descent step. In the case of stochastic gradient descent, the parameter update is executed inside the loop over all examples. This means a gradient descent step is taken each time an example is propagated through the network (cf. Algorithm 2).

Algorithm 1 Training a neural network with full-batch gradient descent. The inner loop is only displayed for a better understanding. Normally, the loop over the examples is vectorized for more efficient computations.

Require: training data X, targets y
  define network architecture (input layer, hidden layers, output layer, activation function) set learning rate α
  initialize weights W and biases b
  for all *epochs* **do**
    for example $i \leftarrow 1$ to m **do**
      apply forward propagation: $\hat{y}_i \leftarrow f_{NN}(x_i; W, b)$ ▷ cf. Sect. 3.2
      compute loss: $C_i \leftarrow (y_i - \hat{y}_i)^2$
      apply backpropagation for gradients $\partial C_i / \partial W$ and $\partial C_i / \partial b$ ▷ cf. Sect. 3.4
    end for
    compute full-batch cost function: $C \leftarrow \frac{1}{m} \sum_{i=1}^{m} C_i$
    compute full-batch gradients w.r.t. W: $\frac{\partial C}{\partial W} \leftarrow \frac{1}{m} \sum_{i=1}^{m} \frac{\partial C_i}{\partial W}$
    compute full-batch gradients w.r.t. b: $\frac{\partial C}{\partial b} \leftarrow \frac{1}{m} \sum_{i=1}^{m} \frac{\partial C_i}{\partial b}$
    update weights: $W \leftarrow W - \alpha \frac{\partial C}{\partial W}$
    update biases: $b \leftarrow b - \alpha \frac{\partial C}{\partial b}$
  end for

Algorithm 2 Training a neural network with stochastic gradient descent.

Require: training data X, targets y
  define network architecture (input layer, hidden layers, output layer, activation function) set learning rate α
  initialize weights W and biases b
  for all *epochs* **do**
    for example $i \leftarrow 1$ to m **do**
      apply forward propagation $\hat{y}_i \leftarrow f_{NN}(x_i; W, b)$ ▷ cf. Sect. 3.2
      compute loss: $C_i \leftarrow (y_i - \hat{y}_i)^2$
      apply backpropagation for gradients $\partial C_i / \partial W$ and $\partial C_i / \partial b$ ▷ cf. Sect. 3.4
      update weights: $W \leftarrow W - \alpha \frac{\partial C_i}{\partial W}$
      update biases: $b \leftarrow b - \alpha \frac{\partial C_i}{\partial b}$
    end for
  end for

The most popular optimization approach in practice is mini-batch gradient descent (cf. Algorithm 3). Instead of computing the gradients averaged over the whole training set, the gradients are evaluated just for a small part of the training data, the so-called mini-batch. This allows to process datasets more efficiently, when they are too big to fit the memory as a whole. Apart from that, mini-batch gradient descent is preferred, because it is said to introduce a regularizing effect [WM03].

Algorithm 3 Training a neural network with mini-batch gradient descent.

Require: training data X, targets y

    define network architecture (input layer, hidden layers, output layer, activation function) set learning rate α

    initialize weights W and biases b

    for all *epochs* **do**

        shuffle rows of X and y synchronously (optional)

        divide X and y into n batches of size k

        for all *batches* **do**

            for example $i \leftarrow 1$ to k **do**

                apply forward propagation: $\hat{y}_i \leftarrow f_{NN}(x_i; W, b)$ $\triangleright$ cf. Sect. 3.2

                compute loss: $C_i \leftarrow (y_i - \hat{y}_i)^2$

                apply backpropagation for gradients $\partial C_i / \partial W$ and $\partial C_i / \partial b$ $\triangleright$ cf. Sect. 3.4

            end for

            compute mini-batch cost function: $C \leftarrow \frac{1}{k} \sum_{i=1}^{k} C_i$

            compute mini-batch gradient w.r.t. W: $\frac{\partial C}{\partial W} \leftarrow \frac{1}{k} \sum_{i=1}^{k} \frac{\partial C_i}{\partial W}$

            compute mini-batch gradients w.r.t. b: $\frac{\partial C}{\partial b} \leftarrow \frac{1}{k} \sum_{i=1}^{k} \frac{\partial C_i}{\partial b}$

            update weights: $W \leftarrow W - \alpha \frac{\partial C}{\partial W}$

            update biases: $b \leftarrow b - \alpha \frac{\partial C}{\partial b}$

        end for

    end for

3.7 Regularization of Neural Networks

As stated in the previous section on regularization (cf. Sect. 2.8), the task of making an algorithm perform well on new inputs and not only on the training data is one of the biggest challenges in machine learning and is a field of extensive research. Regularization includes all strategies aiming to diminish the test error without increasing the training error. Ideally, they trade a significant reduction of variance for a slightly increased bias. There exist multiple approaches to regularize a machine learning model or in particular neural networks. One option is to formulate certain constraints, for example, by directly restricting the parameter values or by adding an extra term to the objective function that constrains the parameters indirectly. Some constraints and restrictions can alter an undetermined problem into a determined one, others prefer simpler models for better generalization properties. The idea to choose the simplest hypothesis among competing explanations stems back to the fourteenth century and is known as Occam's razor. Overall, these constraints and restrictions are often designed to encode prior knowledge about the problem and, if chosen properly, can help to reduce the generalization error [GBC16]. Furthermore, Sect. 2.8 introduced the three cases of underfitting (or high bias), overfitting (or high variance), and an ideal model capacity that matches the complexity of the underlying problem.

Most problems tackled by deep learning algorithms like image recognition or audio sequences are too complex to be modeled precisely. However, experience has shown that building a large model with an appropriate regularization mechanism yields the best results in terms of minimizing the generalization error [GBC16].

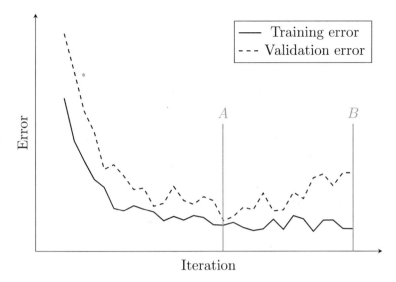

Fig. 3.6 Early stopping: at point A, the validation error reaches its minimum. At point B, the optimization routine is terminated since the validation error has not shown any improvement for the prescribed number of iterations

3.7.1 Early Stopping

When plotting the training and validation error for every iteration step, a common observation is that while the training error is steadily decreasing, the test error starts to rise again at a certain point in time (cf. Fig. 3.6). This usually happens when the model has a sufficient capacity to overfit the problem. The principal idea of early stopping is to halt the training process as soon as the validation error rises and the model enters the overfitting regime (see point A in Fig. 3.6) [Meh+19]. In this way, the fitting of particular features of the training samples can be avoided. It is essential to monitor the error on a validation set and not the test set, because the number of training steps of the gradient-based algorithm has now become a hyperparameter of the model. Every time the validation error decreases, the weights and biases of the model are stored. The algorithm terminates when the validation error has not improved over a predefined number of iteration steps (see point B in Fig. 3.6). Finally, the model parameters at the point of the lowest validation error are returned.

Due to its simplicity and effectiveness, early stopping is a very popular regularization method often applied in practice. It can be seen as a hyperparameter selection algorithm determining the ideal number of training steps. The only additional costs are the evaluations of the validation set after each epoch and the memory used to store the parameters. Conversely, the computational costs are often reduced significantly, since the execution of unnecessary training steps is prevented.

3.7.2 L^1 and L^2 Regularization

The two main representatives from the family of parameter norm penalties have already been introduced in Sect. 2.6, namely L^1 and L^2 regularization. The methods have briefly been described for linear regression, and it is straightforward to extend them to regularize neural networks. Both regularization techniques aim to limit the capacity of a model by penalizing the parameters Θ of the model with the help of a penalty term $\Omega(\Theta)$ that is added to the cost function C [GBC16]. In a general form, this can be expressed as

$$\tilde{C} = C + \lambda\Omega, \tag{3.29}$$

where $\tilde{C}$ is the regularized and C the unregularized cost function. The coefficient λ is a hyperparameter weighting the relative contribution of the penalty term Ω. If $\lambda = 0$, no regularization is applied and larger values for λ result in a more regularized model. In case of L^1 regularization, the cost function takes the form

$$\tilde{C} = C + \lambda||\boldsymbol{w}||_1, \tag{3.30}$$

and

$$\tilde{C} = C + \frac{\lambda}{2}\boldsymbol{w}^T\boldsymbol{w} \tag{3.31}$$

for L^2 regularization, respectively, where C can be any cost function, e.g. the mean squared error.

How these terms influence the training process can be shown when deriving the update rules of the gradient descent algorithm for the regularized cost function $\tilde{C}$ [GBC16]. Beginning with computing the partial derivatives of Eq. (3.30)

$$\frac{\partial\tilde{C}}{\partial\boldsymbol{w}} = \frac{\partial C}{\partial\boldsymbol{w}} + \lambda\,\text{sign}(\boldsymbol{w}), \tag{3.32}$$

where $\text{sign}(\boldsymbol{w})$ is applied element-wise and Eq. (3.31)

$$\frac{\partial\tilde{C}}{\partial\boldsymbol{w}} = \frac{\partial C}{\partial\boldsymbol{w}} + \lambda\boldsymbol{w}, \tag{3.33}$$

the learning rule for the weights takes on the updated form

$$\boldsymbol{w} \rightarrow \boldsymbol{w}' = \boldsymbol{w} - \alpha\lambda\,\text{sign}(\boldsymbol{w}) - \alpha\frac{\partial C}{\partial\boldsymbol{w}} \tag{3.34}$$

for L^1 regularization and

$$w \to w' = w\,(1 - \alpha\lambda) - \alpha\frac{\partial C}{\partial w} \qquad (3.35)$$

for L^2 regularization, respectively. In comparison, the update rule for an unregularized cost function is defined as

$$w \to w' = w - \alpha\frac{\partial C}{\partial w}. \qquad (3.36)$$

Depending on the choice of the norm, different solutions are preferred [Nie15]. L^1 regularization encourages the selection of few high-importance connections, while the other weights are forced toward zero. Looking at Eq. (3.34), the weights always shrink by a constant amount and eventually tend toward zero. By contrast, L^2 regularization shrinks the weights proportional to w, so for small weights, the reduction is much smaller compared to L^1 regularization and, thus, is rather seen as weight decay. One intuitive explanation why smaller weights are preferred is that they reduce the sensitivity toward changes in the inputs. If the weights are large, even a small variation of the input can drastically alter the output.

The given formulas only consider the weights, because regularizing the biases can have unwanted effects like leading to severe underfitting. Theoretically, the hyperparameter λ could be chosen individually for each layer, but usually it is set equally for the whole network to reduce the search space during hyperparameter tuning [GBC16].

Overall, regularization with parameter norm penalties can effectively reduce the generalization error of neural networks and can therefore improve their performance on a multitude of tasks [Nie15]. Nevertheless, the effectiveness is usually demonstrated empirically, rather than with a mathematical proof.

3.7.3 Dropout

In contrast to the parameter norm penalties, dropout regularization alters the network itself instead of modifying the cost function [Nie15]. Before initializing the usual training workflow, half of the hidden neurons get randomly and only temporarily dropped, while input and output neurons stay untouched. Then, forward- and backpropagation as well as the parameter update are executed. Importantly, dropout works with stochastic gradient-based methods in which the cycle is only applied to one mini-batch of the training data (see Sect. 2.6). Afterward, the dropout neurons are restored and the procedure is repeated for the next mini-batch deactivating another random subset of hidden neurons (cf. Fig. 3.7). Thus, the weights and biases are learned with only half of the hidden neurons activated. Conversely, double the amount of hidden neurons is active when the network is eventually used for making predictions. To compensate for that, the outgoing weights of every hidden neuron

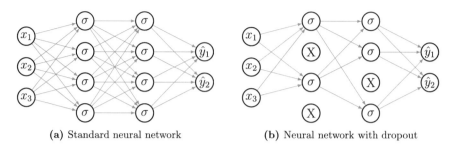

(a) Standard neural network   (b) Neural network with dropout

Fig. 3.7 Dropout. Illustration inspired by Nielsen [Nie15]

are halved. Other quotas than dropping 50% of hidden neurons are also possible and are usually defined with the hyperparameter $p \in [0, 1]$.

In order to understand why dropout helps the model to generalize better, it is worth to look at a related topic known as ensemble method [Nie15]. If several different neural networks are trained with the same training data, they most probably produce different results due to their varying initial states. To select the preferred output, a voting or averaging scheme is applied to the results. Assuming all networks overfit the data in a different way, taking an average over the outputs can help to prevent this type of overfitting. Since training multiple networks is only possible under extensive computational effort, dropout imitates this approach at a much reduced cost. Dropping a set of neurons during each mini-batch update is similar to training different neural networks. The dropout method can be considered as averaging the effects of a large number of various neural networks and thus may reduce overfitting.

3.7.4 Dataset Augmentation

The best way to increase the generalization abilities of a model is to train it on more data. In most cases, obtaining more training data is not feasible, but sometimes the creation of "fake" data can be an option. A good example is the task of image classification, where a high-dimensional input is mapped to a single classifier. This implies that the model has to be invariant to a wide range of transformations. For the example of image recognition, it is fairly easy to introduce small variations like translation, rotation, or scaling to the input images.

Another data augmentation technique is the inclusion of noise [GBC16]. Most classification and regression tasks should still be solvable, even when a small amount of random noise is added to the input data. However, neural networks seem not to be very robust to noisy inputs. A possibility to increase the robustness is to actually train the network with noise-injected data.

3.8 Example: Approximating the Sine Function

As stated in Sect. 3.1, fully connected feed-forward neural networks with at least one hidden layer are capable of approximating any continuous function with arbitrary precision given a sufficient number of hidden neurons. On the example of the sine function, this section investigates some of the introduced characteristics of machine learning and neural networks. The sine is a suitable choice since it is a continuous function and its periodicity offers enough complexity to serve as an academic example. All following results are generated using Python and the PyTorch framework, the latter being a widely used and highly optimized library for training deep neural networks [Pas+19].

The basic architecture of the fully connected feed-forward network is depicted in Fig. 3.8. It consists of one input unit, two hidden layers with 50 hidden neurons each as well as one output unit. The goal is to approximate

$$f(x) = \sin(2\pi x), x \in [-1, 1]$$

in the interval $[-1, 1]$. The neural network approximation can be written as

$$f_{NN} = \boldsymbol{w}_3^T \boldsymbol{\sigma} \left(\boldsymbol{W}_2 \left(\boldsymbol{\sigma}(\boldsymbol{w}_1^T \boldsymbol{x} + \boldsymbol{b}_1) \right) + \boldsymbol{b}_2 \right) + b_3 = \hat{y}, \tag{3.37}$$

with the hyperbolic tangent $\sigma(z_j) = \tanh(z_j)$ serving as the activation function for the hidden neurons.

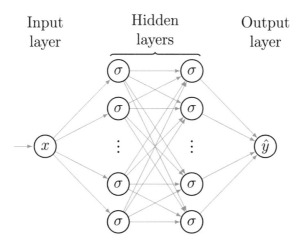

Fig. 3.8 Feed-forward neural network architecture used for approximating the sine function. The network takes x as an input and outputs the corresponding $\hat{y}$ coordinate. Each of the two hidden layers consists of 100 neurons with sigmoid activation functions. The output neuron uses a linear activation

The training set containing 40 samples is generated by computing $y = \sin(2\pi x) + \epsilon$ for a uniform random distribution of values x across the interval $[-1, 1]$. By adding noise with the term ϵ, noisy measurements often occurring in real-world applications are simulated. The noise term is computed as

$$\epsilon = 0.1 \cdot \mathcal{U}(-1, 1),$$

where $\mathcal{U}(-1, 1)$ denotes values drawn from a uniform random distribution. Similarly, a validation set is generated by sampling 40 randomly chosen points in the interval $[-1, 1]$. The test set is simply represented by the analytical solution of the sine function.

Since the prediction of the sine function is a regression task, the mean squared error loss is chosen as the cost function C

$$C = \text{MSE}_{\text{train}} = \frac{1}{m^{(\text{train})}} \sum_{i=1}^{m^{(\text{train})}} (y_i^{(\text{train})} - \hat{y}_i^{(\text{train})})^2. \tag{3.38}$$

Using the cost function C and the respective gradient ∇C, the loss on the training data is minimized by applying full-batch gradient descent, in particular, the commonly used optimizer Adam (in short for: Adaptive Moment Estimation) with a learning rate of $\alpha = 0.001$. The computation of the gradient is handled by PyTorch, which provides a very fast automatic differentiation algorithm to compute the partial derivatives with respect to all parameters of the network.

The weights of the network are initialized by drawing samples from a uniform distribution within certain bounds. The bounds $[-l, l]$ are defined as $l = \sqrt{\frac{1}{n_{in}}}$ with n_{in} being the number of input units in the weight tensor $\boldsymbol{w}^{(l)}$ [Doc20]. The initialization of the biases $\boldsymbol{b}^{(l)}$ follows the same scheme. As already indicated in Sect. 2.6, the choice of initial parameters can have a severe impact on the results since the gradient-based algorithm may descent into a completely different minimum of the cost function. For a detailed and interactive explanation of how the weight initialization influences the convergence of deep neural network training, the reader is referred to the excellent blog post by Katanforoosh and Kunin [KK19].

First results at different epochs of the training process are shown in Fig. 3.9. While the number of training iterations is still low (cf. Fig. 3.9a), the model struggles to fit the training data and the underlying sine function. As soon as the network has learned the parameters for a sufficient amount of time (cf. Fig. 3.9b), it fits the training data and also generalizes well on the test data across the sine. If the training continues for too long (cf. Fig. 3.10a), the model starts to fit every example in the training set. This includes also the noisy data points resulting in overfitting and bad predictions for the test data.

As discussed in Sect. 3.7.2, a way to resolve the problem of overfitting is to use parameter norm penalties like L^2 regularization, which adds a term to the cost function penalizing large weights. The influence of the penalty term is controlled

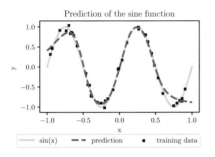
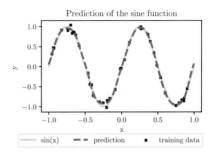

(a) Neural network still showing signs of under-fitting after 500 training epochs.

(b) Neural network properly fitting the underlying sine function after 5000 iterations.

Fig. 3.9 Predictions of the example network at different training epochs

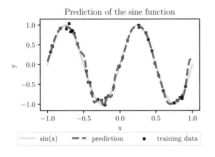
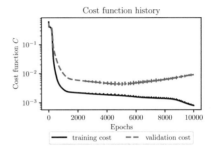

(a) Neural network prediction after 10 000 epochs of training and no regularization. Prediction shows clear signs of overfitting.

(b) Training and validation error plotted for each training iteration. The spikes are caused by the Adam optimizer adapting the learning rate.

Fig. 3.10 The example network trying to predict the sine function after 10 000 training iterations

by the hyperparameter λ. The plots in Fig. 3.11 show the results with applied L^2 regularization for different values of λ. For large values of λ, the model exhibits underfitting, meaning it fails to fit the training data as well as the test data (cf. Fig. 3.11b). In contrast, when the ideal value for λ is found, the network is able to fit the data points of the training set, and also a generalization to the test data is achieved (cf. Fig. 3.11a). If no regularization ($\lambda = 0$) is applied, the model shows the typical signs of overfitting (cf. Fig. 3.10a). So, the function exactly passes through almost every noisy data point in the training data, but fails to represent the underlying sine wave.

Investigating the seemingly simple example of approximating the sine function reveals that the results are sensitive to a great variety of factors. All the different kinds of hyperparameters and possible settings determined by the user have a great influence on the output and the learning speed of the model. For example, choosing a different activation function, a different learning rate, or even another optimizer can

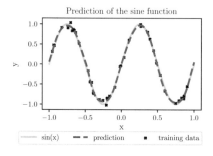

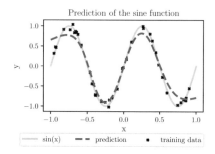

(a) Prediction after 10 000 epochs of training and L^2 regularization ($\lambda = 1 \times 10^{-5}$).

(b) Prediction after 10 000 epochs of training and L^2 regularization ($\lambda = 1 \times 10^{-2}$).

Fig. 3.11 Regularization of sine example

drastically change the outcome. Figure 3.12 displays a few representative results that demonstrate the influence of the three aforementioned settings and hyperparameters.

3.9   Input Derivatives

Chapters 5 and 6 introduce neural networks that approximate solutions to typical physics and engineering problems governed by partial differential equations. The neural networks predict physical quantities of interest from a set of independent input variables such as space or time. Since neural networks are fully differentiable functions, it is not only possible to compute the derivatives with respect to the network parameters, but also with respect to the input variables. Even higher-order derivatives can be evaluated and used to check if the solution satisfies the governing partial differential equation of the problem.

Computing the derivatives of the network output with respect to the input variables is demonstrated on the example from Sect. 3.2 (cf. Fig. 3.5). To this end, we recall the corresponding analytic expression of the network

$$\hat{y} = \sigma(w_1^{(2)}\sigma(w_{11}^{(1)}x_1 + w_{12}^{(1)}x_2) + w_2^{(2)}\sigma(w_{21}^{(1)}x_1 + w_{22}^{(1)}x_2)). \tag{3.39}$$

For instance, the derivative of the network output $\hat{y}_1$ with respect to the input variable x_1 takes on the following form:

$$\begin{aligned}
\frac{\partial \hat{y}_1}{\partial x_1} &= \sigma'(w_1^{(2)}\sigma(w_{11}^{(1)}x_1 + w_{12}^{(1)}x_2) + w_2^{(2)}\sigma(w_{21}^{(1)}x_1 + w_{22}^{(1)}x_2)) \\
&\quad \cdot (w_1^{(2)}\sigma'(w_{11}^{(1)}x_1 + w_{12}^{(1)}x_2)w_{11}^{(1)} + w_2^{(2)}\sigma'(w_{21}^{(1)}x_1 + w_{22}^{(1)}x_2)w_{21}^{(1)}).
\end{aligned} \tag{3.40}$$

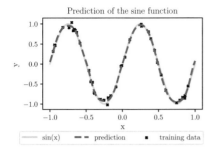

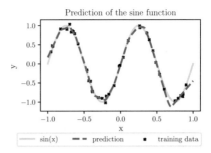

(a) Reference neural network with sigmoid activations properly fitting the underlying sine function after 5000 iterations. The network parameters are learned by applying the Adam optimizer with a learning rate of $\alpha = 0.001$.

(b) Prediction generated using ReLU as the activation function. With ReLU activations the network already starts to overfit at 5000 epochs.

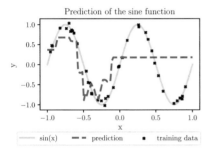

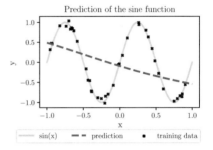

(c) Prediction after 5000 epochs with learning rate $\alpha = 0.1$. This example shows that a large learning rate can prevent the network from learning.

(d) Prediction after 5000 epochs using stochastic gradient descent (SGD) instead of the Adam optimizer. A drawback of SGD is the slower convergence rate, since the parameters are updated after the evaluation of each single example (cf. Section 3.6).

Fig. 3.12 Parameter study for the example network predicting the sine function. Besides the parameter in question, all parameters remain the same as in the reference example on the top left

As for the derivatives with respect to the network parameters, an analytical differentiation is not feasible in practice. Here, deep learning frameworks that make use of automatic differentiation such as TensorFlow or PyTorch are equally suited for a fast and precise numerical evaluation of the desired derivatives (cf. Sect. 3.3).

3.10 Advanced Architectures

The following sections introduce two network architectures that have been very successful in their respective field of application. Adapted to a particular data structure, they achieve superior results in various tasks when compared to fully connected feed-forward neural networks.

3.10.1 Convolutional Neural Network

Convolutional neural networks (CNNs) were originally designed for image processing [Lec+98]. Their distinct architecture allows them to account for the spatial structure of a picture and thus makes convolutional networks the preferred choice for image classification tasks [Nie15]. When recognizing an object in an image, it can be assumed that features describing the object are found in close proximity to each other. Furthermore, the exact location in the image is not of importance for the identification of an object, just the relative position of features to each other matters. CNNs exploit these two realities about physical objects, namely locality and translational variance. Like an ordinary neural network, a CNN consists of neurons that have learnable weights and biases. What differs is the arrangement of the processed data and the mathematical operation employed in the hidden layers, which are separated into convolutional and pooling operations. The input layer is no longer a vector but has the shape of a matrix to account for the grid-like structure of pixels in an image. For instance, a convolutional network is processing a square image of 7×7 pixels where each pixel corresponds to a gray-scale value. Then, a hidden neuron in the convolutional layer is connected with a small region of the input pixels, e.g. a 3×3 window, called the local receptive field of the hidden neuron (see Fig. 3.13). The size of the local receptive field and the distance between the fields, also referred to as stride length, define the number of neurons of the convolutional layer. In the example shown in Fig. 3.13, the local receptive field of size 3×3 moves over the input layer with a stride length of one. Since the input layer consists of 7×7 pixels, the corresponding convolutional layer requires 5×5 neurons. Together with the size of the local receptive field, the stride length is a hyperparameter of the convolutional network. A specialty of CNNs is that all hidden neurons in the convolutional layer share the same weights and bias, often referred to as the filter K. As a result, all neurons detect the same feature at different locations in the input image, which implies the characteristic of translational invariance. The output of the convolution operation is called the feature map. A convolutional layer normally consists of more than one feature map in order to detect multiple localized features. Every feature map is generally followed by the application of a pooling operation that condenses the output of the convolution (see Fig. 3.14). For example, each unit in the pooling layer summarizes a region of 2×2 neurons in the previous layer. A common pooling technique is max-pooling, where simply the maximum value of the respective region is

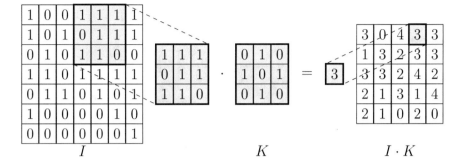

Fig. 3.13 Convolution layer: the local receptive field or filter K of size 3×3 with a stride length of one is applied to the input layer I

Fig. 3.14 Max-pooling with a 2×2 filter and stride length of one ($\max\{1, 2, 1, 2\} = 2$)

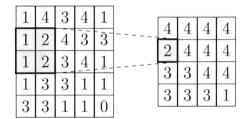

taken. Pooling can be interpreted as a query that checks if a feature has been detected somewhere while neglecting the exact position in the image. In the end, a fully connected layer is added to perform the classification task. The arrangement of multiple convolutional and pooling layers allows very deep and expressive network architectures. Sharing weights and biases in combination with the pooling operation reduces the amount of trainable parameters significantly. Thus, convolutional networks learn much faster than fully connected neural networks with comparable expressive power [Nie15]. The backpropagation algorithm for calculating the gradients only needs minor adaptations to work with the convolution and pooling operations.

3.10.2 Recurrent Neural Network

Conventional neural networks are just able to evaluate a current state described by the fixed-size input vector. So information from previous states cannot persist nor be passed on to later ones. Recurrent neural networks (RNNs) address this issue by adding loops that allow the persistence and propagation of information over time. A recurrent neural network can be seen as multiple copies of the same network, each forwarding a message to their respective successor [Ola15b]. The sketch of an unrolled recurrent neural network is depicted in Fig. 3.15. As shown, RNNs are

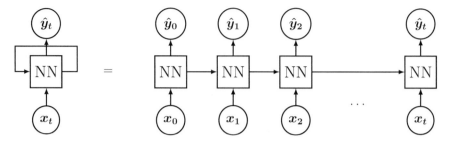

Fig. 3.15 An unrolled recurrent neural network. Adapted from [Ola15b] with permission

chain-like structures that take in sequences of data $x_0, x_1, \ldots, x_t$ and generate a sequential output $\hat{y}_0, \hat{y}_1, \ldots, \hat{y}_t$. Sequential data often occurs in tasks like speech recognition, language modeling, and translation, making recurrent neural networks the preferred choice for these kinds of problems. One shortcoming of standard RNNs is that they are only able to connect very recent information with the current task. An illustrative example is the prediction of words in text sequences. For instance, the final word in the sentence "the color of a lemon is *yellow*" can directly be derived from the preceding words. Often the relevant information needed for the prediction is found much earlier in the sequence. In the text excerpt "We used too many lemons for our lemonade... We did not like the lemonade, it was too *sour*", the gap between the information and the place of prediction is much larger and RNNs usually fail to learn such long-term dependencies [Hoc91]. This problem can be solved by introducing a special implementation of recurrent neural networks, called long short-term memory networks (LSTMs) that often achieve outstanding results in the aforementioned tasks. The key concept of LSTMs is their cell state, where information has to pass several gates before it gets forwarded. Three different gates control and protect the cell state by deciding if the information is added, kept, or discarded. In this way, LSTMs are capable to remember information and to store long-term dependencies. Since recurrent neural networks employ a sequential structure, the backpropagation algorithm is adapted to take into account the temporal component resulting in a method called "backpropagation through time" [Ola15b].

For more in-depth content about recurrent neural networks, the reader is referred to Chap. 10 of Goodfellow's "Deep Learning" book [GBC16, Chap. 10] and Christopher Olah's blog post "Understanding LSTMs" [Ola15b].

References

[Mar19]   Florian Marquardt. "Machine Learning for Physicists". Lecture. Lecture. 2019. URL: https://www.video.uni-erlangen.de/clip/id/10611.html (visited on 03/30/2020).

[Dem+14] Howard B. Demuth et al. *Neural Network Design*. 2nd. Stillwater, OK, USA: Martin Hagan, 2014. ISBN: 978-0-9717321-1-7.

[Car+19]   Giuseppe Carleo et al. "Machine learning and the physical sciences". In: *Rev. Mod. Phys.* 91.4 (Dec. 6, 2019), p. 045002. ISSN: 0034-6861, 1539-0756. DOI https://doi.org/10.1103/RevModPhys.91.045002. URL: http://arxiv.org/abs/1903.10563 (visited on 01/15/2020).

[Meh+19]   Pankaj Mehta et al. "A high-bias, low-variance introduction to Machine Learning for physicists". In: *Physics Reports* 810 (May 2019), pp. 1–124. ISSN: 03701573. DOI https://doi.org/10.1016/j.physrep.2019.03.001. URL: http://arxiv.org/abs/1803.08823 (visited on 01/14/2020).

[GBC16]   Ian Goodfellow, Yoshua Bengio, and Aaron Courville. *Deep Learning*. MIT Press, 2016. ISBN: 0-262-03561-8. URL: http://www.deeplearningbook.org.

[Cyb89]   G. Cybenko. "Approximation by superpositions of a sigmoidal function". In: *Math. Control Signal Systems* 2.4 (Dec. 1, 1989), pp. 303–314. ISSN: 1435-568X. DOI https://doi.org/10.1007/BF02551274 (visited on 05/20/2020).

[Nie15]   Michael A. Nielsen. *Neural Networks and Deep Learning*. Determination Press, 2015. URL: http://neuralnetworksanddeeplearning.com (visited on 03/13/2020).

[MLP16]   Hrushikesh Mhaskar, Qianli Liao, and Tomaso Poggio. "Learning Functions: When Is Deep Better Than Shallow". In: arXiv:1603.00988 [cs] (May 29, 2016) (visited on 05/20/2020).

[HH79]   Edwin Hewitt and Robert E. Hewitt. "The Gibbs-Wilbraham phenomenon: An episode in fourier analysis". In: *Arch. Hist. Exact Sci.* 21.2 (1979), pp. 129–160. ISSN: 0003-9519, 1432-0657. DOI https://doi.org/10.1007/BF00330404 (visited on 11/05/2020).

[LLS08]   B. Llanas, S. Lantarón, and F. J. Sáinz. "Constructive Approximation of Discontinuous Functions by Neural Networks". In: *Neural Process Lett* 27.3 (June 2008), pp. 209–226. ISSN: 1370-4621, 1573-773X. DOI https://doi.org/10.1007/s11063-007-9070-9 (visited on 05/20/2020).

[Mar17]   Florian Marquardt. *Visualize the output of a multilayer network*. 2017. URL: http://www.thp2.nat.uni-erlangen.de/images/0/0e/MultiLayerNet_SimpleExample.py (visited on 07/29/2020).

[Bay+18]   Atılım Gunes Baydin et al. "Automatic Differentiation in Machine Learning: a Survey". In: (2018), p. 43.

[Ola15a]   Christopher Olah. *Calculus on Computational Graphs: Backpropagation*. colah's blog. Aug. 31, 2015. URL: http://colah.github.io/posts/2015-08-Backprop/ (visited on 10/01/2020).

[Ng20]   Andrew Ng. "Machine Learning". Online course. Online course. 2020. URL: https://www.coursera.org/learn/machine-learning?.

[Cho18]   François Chollet. *Deep learning with Python*. OCLC: ocn982650571. Shelter Island, New York: Manning Publications Co, 2018. 361 pp. ISBN: 978-1-61729-443-3.

[WM03]   D. Randall Wilson and Tony R. Martinez. "The general inefficiency of batch training for gradient descent learning". In: *Neural Networks* 16.10 (Dec. 2003), pp. 1429–1451. ISSN: 08936080. DOI https://doi.org/10.1016/S0893-6080(03)00138-2 (visited on 07/11/2020).

[Pas+19]   Adam Paszke et al. "PyTorch: An Imperative Style, High-Performance Deep Learning Library". In: arXiv:1912.01703 [cs, stat] (Dec. 3, 2019) (visited on 09/29/2020).

[Doc20]   PyTorch Documentation. *Linear — PyTorch 1.6.0 documentation*. 2020. URL: https://pytorch.org/docs/ (visited on 09/29/2020).

[KK19]   Kian Katanforoosh and Daniel Kunin. *Initializing neural networks*. deeplearning.ai. 2019. URL: https://www.deeplearning.ai/ai-notes/initialization/ (visited on 12/22/2020).

[Lec+98]   Y. Lecun et al. "Gradient-based learning applied to document recognition". In: *Proceedings of the IEEE* 86.11 (Nov. 1998). Conference Name: Proceedings of the IEEE, pp. 2278–2324. ISSN: 1558-2256. DOI https://doi.org/10.1109/5.726791.

[Ola15b]   Christopher Olah. *Understanding LSTM Networks*. colah's blog. Aug. 27, 2015. URL: http://colah.github.io/posts/2015-08-Understanding-LSTMs/ (visited on 04/29/2020).

[Hoc91]   Sepp Hochreiter. "Untersuchungen zu dynamischen neuronalen Netzen". In: *Diploma, Technische Universität München* 91.1 (1991).

Chapter 4
Machine Learning in Physics and Engineering

The access to enormous quantities of data combined with rapid advances in machine learning in recent years yielded outstanding results in the fields of computer vision, recommendation systems, medical diagnosis, or financial forecasting [AML12]. Nonetheless, the impact of learning algorithms reaches far beyond and already found its way into many scientific disciplines [Adi+18]. While machine learning frameworks are already able to support radiologists in medical diagnostics [Lam+12], scientists from other fields only begin to explore the immense potential of data-driven algorithms.

Even before the unprecedented success of deep learning, a handful of academics identified the potential of neural networks in scientific computations during the 1990s and early 2000s. For instance, Rico-Martínez and Kevrekidis used a neural-network-based approach for the identification of non-linear systems that are continuous in time [RK93]. With the help of a neural network, Milano and Koumoutsakos demonstrated the reconstruction of a near-wall field in turbulent flow [MK02]. An example from the field of chemical engineering is the process modeling of a fed-batch bioreactor. Psichogios and Ungar described a hybrid network architecture that incorporates additional information about the problem [PU92]. The idea of enriching a neural network architecture with prior knowledge is also found in the work of Lagaris et al., who proposed an artificial neural network for solving ordinary and partial differential equations [LLF98]. The last two examples inspired the paper by Raissi et al. [RPK19], which is discussed in more detail in Chap. 5.

4.1 General Reviews

Solving scientific problems with the help of machine learning techniques is a very broad and active area of research. The following section gives a short introduction to the topic and provides the reader with an overview of current literature.

© The Author(s), under exclusive license to Springer Nature Switzerland AG 2021
S. Kollmannsberger et al., *Deep Learning in Computational Mechanics*,
Studies in Computational Intelligence 977,
https://doi.org/10.1007/978-3-030-76587-3_4

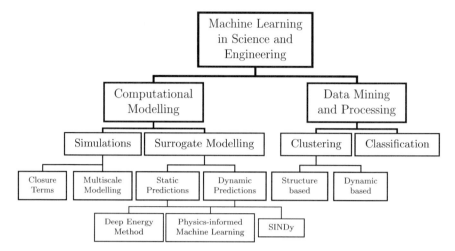

Fig. 4.1 Machine Learning in Science and Engineering. The figure is adapted from Frank et al. [FDC20] and extended under the CC BY 4.0 License

In order to validate existing theories or new hypotheses, scientists heavily rely on empirical studies. This means a great part of scientific work is dedicated to data analysis. For instance, the sub-field of statistical physics deals with the derivation of physical models from experimental data. According to Mehta et al., statistical physics shares a lot of common ground with statistical learning theory, the fundamental concept of learning probability distributions from data [Meh+19]. Their recommendable introduction to the world of algorithmic data analysis is addressed to interested readers with a background in physical sciences. Carleo et al. take on a similar perspective [Car+19]. They discuss the interface between machine learning and physics and further highlight how both fields could equally benefit from each other. According to the authors, one example of this potential symbiosis is quantum computing. Machine learning can support the building process and analysis of quantum computers. On the other hand, the execution of learning algorithms on quantum computers could lead to a significant speed-up of the training process.

A review article about the applications of machine learning in natural sciences by Frank et al. puts a greater emphasis on recent developments within computational sciences and engineering [FDC20]. Next to scientific data analysis supported by well-established learning algorithms, they discuss machine learning in the context of computational simulation and modeling. More specifically, they distinguish between algorithms that help to improve conventional computational methods and different classes of surrogate models that can replace those methods completely in specific situations (cf. Fig. 4.1).

Conventional methods in the context of computational mechanics usually refer to the finite element method (FEM) or other discretization methods such as the finite difference or finite volume method. The core idea of discretization methods is to subdivide a large domain into simpler and smaller parts that can be easily processed by a computer. These methods allow the numerical solution of partial differential equations and they are applied to a variety of static and dynamic problems. Due to its robustness, the finite element method is the most popular approach for solving problems in the fields of structural analysis, heat transfer, multi-physics applications as well as fluid mechanics [All07]. The application of finite differences and finite volumes is more prevalent in problems of thermodynamics and computational fluid dynamics [PTA12].

Time-dependent partial differential equations are of major significance in the field of fluid dynamics. Fluids in motion often exhibit highly non-linear behavior, which makes their solution with conventional numerical methods very costly. In the opinion of Brunton et al., fluid mechanics could benefit from the application of machine learning due to its ability to cope with non-linear relations [BNK20]. Their review treats recent applications in problems of flow modeling and optimization as well as experimental flow control. While highlighting the success of machine learning in critical tasks like model order reduction or feature extraction, they pointed out current obstacles that demand future research.

Inspired by recent results in reinforcement learning (cf. Sect. 2.3.4), Garnier et al. assessed the first approaches that apply deep reinforcement learning in the context of fluid dynamics [Gar+19]. On the examples of flow control and shape optimization, new ideas are compared to classical methods.

Computationally challenging tasks in mechanics do not only arise in structural or fluid dynamics. Also, the domain of material mechanics could possibly benefit from the introduction of machine learning techniques. A collection of articles published by Huber et al. presents data-driven approaches contributing to the advancement of continuum material mechanics [Hub+20].

The cited reviews agree on the great potential of machine learning in applications of physical sciences and engineering. No less could machine learning profit from a wider adoption in the scientific community. For instance, the interface between fluid mechanics and machine learning raises hope for a fruitful exchange of ideas between the two fields [BNK20]. However, some authors noted that machine learning introduces new uncertainties and drawbacks. Data-driven algorithms often lack robustness and cannot guarantee convergence [Gar+19]. Further, the interpretability and explainability of results are compromised, when algorithms are simply applied as "black-boxes" [BNK20].

4.2 Combined Methods

After a more general introduction on potential uses of machine learning in science and engineering, this section summarizes some attempts that augment existing methods for modeling and simulation of physical problems.

Solving partial differential equations while maintaining small-scale features of the solution becomes computationally infeasible for large time scales. In these cases, equations that represent a coarse-grid approximation of the underlying problem are derived. However, it is not always possible to find such a suitable approximation function analytically. Motivated by this circumstance, Ben-Sinai et al. proposed a neural network that learns an effective approximation from actual solutions of the underlying partial differential equations [Bar+19]. This data-driven approach allows accurate results for a much coarser time discretization when compared to the standard finite differences method. Nevertheless, the computational overhead due to the convolutional operations of the neural network is much higher. Additionally, the scalability needs further investigation, since the method was only demonstrated on examples with one spatial dimension.

The finite element method requires the element-wise calculation of integrals [All07]. A standard approach used for the numerical integration is the Gaussian quadrature, which approximates an integral with a finite sum [Ise08]. Oishi et al. proposed a method that optimizes the quadrature rule for computation of the finite element stiffness matrix using a deep neural network [OY17]. The resulting quadrature rule was more accurate than the standard Gauss–Legendre quadrature for the same amount of integration points. Improving the performance of numerical integration is of general importance to computational mechanics, not only in the context of the finite element method. Unfortunately, the computational costs of training a neural network for each problem individually has not been demonstrated to justify the gain in accuracy yet.

Material modeling is an essential part of simulating physical entities. The material models are usually derived from experimental data and are then calibrated for further calculations. According to Kirchdoerfer et al., the additional step of empirical material modeling is a non-negligible source of error to the solution of complex systems [KO16]. Their work introduced a method that replaces the empirical material modeling process with data-driven computations. The proposed solver directly utilizes experimental material data in combination with essential constraints and conservation laws. The first results were conducted on the examples of non-linear three-dimensional trusses and linear elastic solids showing good convergence properties. Even though their approach was formulated in the context of quasi-static mechanics, the authors believe that an extension to dynamic problems is possible.

4.3 Surrogate Models

The previously discussed examples followed the idea of improving existing numerical methods by means of data-driven algorithms. A different approach to exploit the predictive power of machine learning algorithms is to use them as surrogate models for physical simulations. The following publications introduced the first attempts of replacing costly and time-consuming computations with data-driven predictions.

Image-guided interventions are exemplary for clinical applications that demand immediate feedback from practitioners. Due to the high complexity of biomechanical models, standard numerical methods, such as the finite element method, are in many cases not suitable for the application in time-sensitive tasks. On the way to providing real-time results on patient-specific geometries, Liang et al. proposed a deep learning framework that can directly estimate the stress distribution in the wall of an aorta [Lia+18]. The stress distributions used to train the deep neural network were generated by finite element analysis of 729 patient-specific geometries. Martínez-Martínez et al. provided another example from the field of medical applications [Mar+17]. In contrast to the preceding approach, they made use of tree-based methods to simulate the biomechanical behavior of breast tissues during image-guided interventions. Again, the training data was based on finite element simulations of ten real breast models.

In the early design phases of engineering structural components, it is crucial to run multiple iterations, e.g. for the shape optimization of an airfoil. When the problem involves fluid dynamics, the computation in each iteration is costly. Even with efficient solvers, the engineer's workflow is delayed by long waiting times since the solution has to be computed for each and every change in design. The concept of Afshar et al. is to estimate the pressure field and velocity of dynamic problems using a convolutional neural network (cf. Sect. 3.10.1) that learns from pixelated solutions [Afs+19]. In particular, they demonstrated the prediction of a two-dimensional flow field around different airfoil geometries under variable flow conditions and compared the performance to a classical Reynolds-averaged Navier–Stokes solver. The accuracy of the results was sufficient for the early design stage, while the computational time decreased by a factor of four on the available hardware. Three years earlier, Guo et al. proposed a very similar idea [GLI16]. Instead of computing the solution of airflow around an obstacle with conventional methods, they trained a convolutional neural network to predict the resulting velocity field. The experiments were conducted on a greater variety of shapes compared to the first approach, including simple three-dimensional geometries. In terms of speed-up, they claimed to accelerate the computations by two orders of magnitudes in comparison to classical numerical methods.

Motivated by the goal of simulating physics in real time, progress is made in the field of computer graphics. Wiewel et al. introduced an interesting data-driven framework for the prediction of fluid flows [WBT19]. In a first step, they generated datasets with a classical Navier–Stokes solver. Then, a convolutional neural network (CNN) was trained to learn a mapping from the three-dimensional problem into a smaller spatial representation. At the same time, the network learned the corresponding inverse mapping. The reduced model was then fed to an LSTM neural network (cf. Sect. 3.10.2) that predicted the temporal evolution in the reduced space. Finally, the earlier learned reverse mapping transformed the output of the LSTM network back into the three-dimensional space. The whole process is also depicted in Fig. 4.2 [WBT19]. Due to the efficient compression of the CNN, this method allows significant speed-ups compared to conventional fluid flow simulations according to the authors. Furthermore, the proposed work shows good generalization capabilities. From an engineering viewpoint, it is important to note that the accuracy of the results is mainly judged on visual comparison to the reference computation.

All presented works demonstrate the possibility of replacing physically complex simulations with data-driven computations. Using a learned surrogate model

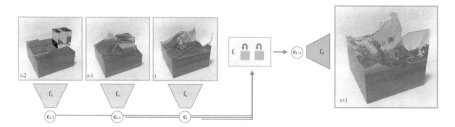

Fig. 4.2 A combination of long short-term memory and convolutional neural network for fluid flow prediction. Here, the trapezoidal shapes denote the encoding and decoding of spatial information by the convolutional layers and the rectangle in the middle represents the LSTM network predicting the temporal evolution in the reduced space. Figure by Wiewel et al. [WBT19], figure reprinted with permission from [WBT21]

may lead to a significant speed-up, which enables the online application in time-sensitive tasks. As a consequence, the problem of expensive data generation and time-consuming training shifts to computations which are carried out in an offline phase i.e. prior to the analysis of the concrete model. A major drawback of this approach is the limited generalization ability. The machine learning frameworks are trained for one specific task and cannot be used for arbitrary problems. Nevertheless, such approaches can be very efficient. Their basic ideas stem from the more classical field of surrogate modeling which has gained a lot of traction in the light of data driven science and engineering. The reader is referred to Brunton and Kutz for an excellent introduction into this relevant and rapidly developing topic [BK17].

References

[AML12]   Yaser S. Abu-Mostafa, Malik Magdon-Ismail, and Hsuan-Tien Lin. *Learning From Data*. S.l.: AMLBook, 2012. 213 pp. ISBN: 978-1-60049-006-4.

[Adi+18]   Jeff Adie et al. "Deep Learning for Computational Science and Engineering". In: GPU Technology Conference. 2018.

[Lam+12]   Philippe Lambin et al. "Radiomics: Extracting more information from medical images using advanced feature analysis". In: *European Journal of Cancer* 48.4 (Mar. 2012), pp. 441–446. ISSN: 09598049. https://doi.org/10.1016/j.ejca.2011.11.036. URL: https://linkinghub.elsevier.com/retrieve/pii/S0959804911009993 (visited on 07/02/2020).

[RK93]   R. Rico-Martinez and I.G. Kevrekidis. "Continuous time modeling of nonlinear systems: a neural network-based approach". In: *IEEE International Conference on Neural Networks*. IEEE International Conference on Neural Networks. San Francisco, CA, USA: IEEE, 1993, pp. 1522–1525. ISBN: 978-0-7803-0999-9. https://doi.org/10.1109/ICNN.1993.298782. URL: http://ieeexplore.ieee.org/document/298782/ (visited on 07/02/2020).

[MK02]   Michele Milano and Petros Koumoutsakos. "Neural Network Modeling for Near Wall Turbulent Flow". In: *Journal of Computational Physics* 182.1(Oct. 2002), pp. 1–26. ISSN: 00219991. https://doi.org/10.1006/jcph.2002.7146. URL: https://linkinghub.elsevier.com/retrieve/pii/S0021999102971469 (visited on 07/02/2020).

[PU92]   Dimitris C. Psichogios and Lyle H. Ungar. "A hybrid neural network-rst principles
           approach to process modeling". In: *AIChE J.* 38.10 (Oct. 1992), pp. 1499–1511. ISSN:
           0001-1541, 1547-5905. https://doi.org/10.1002/aic.690381003. URL: http://doi.wiley.
           com/10.1002/aic.690381003 (visited on 07/02/2020).

[LLF98]   I.E. Lagaris, A. Likas, and D.I. Fotiadis. "Articial neural networks for solving ordi-
           nary and partial di erential equations". In: *IEEE Trans. Neural Netw.* 9.5 (Sept.
           1998), pp. 987–1000. ISSN: 10459227. https://doi.org/10.1109/72.712178. URL:
           http://ieeexplore.ieee.org/document/712178/ (visited on 01/08/2020).

[RPK19]   M. Raissi, P. Perdikaris, and G.E. Karniadakis. "Physics-informed neural networks: A
           deep learning framework for solving forward and inverse problems involving nonlinear
           partial di erential equations". In: *Journal of Computational Physics* 378 (Feb. 2019), pp.
           686–707. ISSN: 00219991. https://doi.org/10.1016/j.jcp.2018.10.045. URL: https://
           linkinghub.elsevier.com/retrieve/pii/S0021999118307125 (visited on 01/08/2020).

[Meh+19]   Pankaj Mehta et al. "A high-bias, low-variance introduction to Machine Learning for
           physicists". In: Physics Reports 810 (May 2019), pp. 1–124. ISSN: 03701573. https://
           doi.org/10.1016/j.physrep.2019.03.001. arXiv:1803.08823. URL: http://arxiv.org/abs/
           1803.08823 (visited on 01/14/2020).

[Car+19]   Giuseppe Carleo et al. "Machine learning and the physical sciences". In: *Rev. Mod.
           Phys.* 91.4 (Dec. 6, 2019), p. 045002. ISSN: 0034-6861, 1539-0756. https://doi.org/
           10.1103/RevModPhys.91.045002. arXiv:1903.10563. URL: http://arxiv.org/abs/1903.
           10563 (visited on 01/15/2020).

[FDC20]   Michael Frank, Dimitris Drikakis, and Vassilis Charissis. "Machine-Learning Methods
           for Computational Science and Engineering". In: *Computation* 8.1 (Mar. 3, 2020),
           p. 15. ISSN: 2079-3197. https://doi.org/10.3390/computation8010015. URL: https://
           www.mdpi.com/2079-3197/8/1/15 (visited on 07/02/2020).

[All07]   Grégoire Allaire. *Numerical analysis and optimization: an introduction to mathematical
           modelling and numerical simulation.* Numerical mathematics and scientific computa-
           tion. OCLC: ocm82671667. Oxford ; New York: Oxford University Press, 2007. 455
           pp. ISBN: 978-0-19-920521-9.

[PTA12]   Richard H. Pletcher, John C. Tannehill, and Dale Anderson. *Computational Fluid
           Mechanics and Heat Transfer, Third Edition.* Google-Books- ID: Cv4IERczJ4oC. CRC
           Press, Aug. 30, 2012. 777 pp. ISBN: 978-1-59169-037-5.

[BNK20]   Steven Brunton, Bernd Noack, and Petros Koumoutsakos. "Machine Learning for
           Fluid Mechanics". In: *Annu. Rev. Fluid Mech.* 52.1 (Jan. 5, 2020), pp. 477–508.
           ISSN: 0066-4189, 1545-4479. https://doi.org/10.1146/annurev-fluid-010719-060214.
           arXiv:1905.11075. URL: http://arxiv.org/abs/1905.11075 (visited on 06/26/2020).

[Gar+19]   Paul Garnier et al. "A review on Deep Reinforcement Learning for Fluid Mechanics".
           In: arXiv:1908.04127 *[physics]* (Aug. 12, 2019). arXiv: 1908.04127. URL: http://arxiv.
           org/abs/1908.04127 (visited on 06/29/2020).

[Hub+20]   Norbert Huber et al., eds. *Machine Learning and Data Mining in Materials Science.*
           Frontiers Research Topics. Frontiers Media SA, 2020. ISBN: 978-2-88963-651-
           8. https://doi.org/10.3389/978-2-88963-651-8. URL: https://www.frontiersin.org/
           research-topics/8312/machine-learning-and-data-mining-in-materials-science   (vis-
           ited on 07/04/2020).

[Bar+19]   Yohai Bar-Sinai et al. "Learning data-driven discretizations for partial differential equa-
           tions". In: *Proc Natl Acad Sci USA* 116.31 (July 30, 2019), pp. 15344–15349. ISSN:
           0027-8424, 1091-6490. https://doi.org/10.1073/pnas.1814058116. URL: http://www.
           pnas.org/lookup/doi/10.1073/pnas.1814058116 (visited on 01/08/2020).

[Ise08]   Arieh Iserles. *A First Course in the Numerical Analysis of Differential Equations.*
           Google-Books-ID: 3acgAwAAQBAJ. Cambridge University Press, Nov. 27, 2008. 481
           pp. ISBN: 978-1-139-47376-7.

[OY17]   Atsuya Oishi and Genki Yagawa. "Computational mechanics enhanced by deep learn-
           ing". In: *Computer Methods in Applied Mechanics and Engineering* 327 (Dec. 2017), pp.

327–351. ISSN: 00457825. https://doi.org/10.1016/j.cma.2017.08.040. URL: https://linkinghub.elsevier.com/retrieve/pii/S0045782517306199 (visited on 07/01/2020).

[KO16]   Trenton Kirchdoerfer and Michael Ortiz. "Data-driven computational mechanics". In: *Computer Methods in Applied Mechanics and Engineering* 304 (June 2016), pp. 81–101. ISSN: 00457825. https://doi.org/10.1016/j.cma.2016.02.001. arXiv: 1510.04232. URL: http://arxiv.org/abs/1510.04232 (visited on 06/30/2020).

[Lia+18] Liang Liang et al. "A deep learning approach to estimate stress distribution: a fast and accurate surrogate of nite-element analysis". In: *J. R. Soc. Interface* 15.138 (Jan. 31, 2018), p. 20170844. ISSN: 1742-5689, 1742-5662. https://doi.org/10.1098/rsif.2017.0844. URL: https://royalsocietypublishing.org/doi/10.1098/rsif.2017.0844 (visited on 01/14/2020).

[Mar+17] F. Martínez-Martínez et al. "A nite element-based machine learning approach for modeling the mechanical behavior of the breast tissues under compression in real-time". In: *Computers in Biology and Medicine* 90 (Nov. 2017), pp. 116–124. ISSN: 00104825. https://doi.org/10.1016/j.compbiomed.2017.09.019. URL: https://linkinghub.elsevier.com/retrieve/pii/S0010482517303177 (visited on 07/08/2020).

[Afs+19] Yaser Afshar et al. "Prediction of Aerodynamic Flow Fields Using Convolutional Neural Networks". In: *Comput Mech* 64.2 (Aug. 2019), pp. 525–545. ISSN: 0178-7675, 1432-0924. https://doi.org/10.1007/s00466-019-01740-0. arXiv: 1905.13166. URL: http://arxiv.org/abs/1905.13166 (visited on 07/08/2020).

[GLI16]   Xiaoxiao Guo, Wei Li, and Francesco Iorio. "Convolutional Neural Networks for Steady Flow Approximation". In: *Proceedings of the 22nd ACM SIGKDD International Conference on Knowledge Discovery and Data Mining*. KDD'16: The 22nd ACM SIGKDD International Conference on Knowledge Discovery and Data Mining. San Francisco California USA: ACM, Aug. 13, 2016, pp. 481–490. ISBN: 978-1-4503-4232-2. https://doi.org/10.1145/2939672.2939738. URL: https://dl.acm.org/doi/10.1145/2939672.2939738 (visited on 07/01/2020).

[WBT19]   Steffen Wiewel, Moritz Becher, and Nils Thuerey. "Latent-space Physics: Towards Learning the Temporal Evolution of Fluid Flow". In: arXiv:1802.10123 [cs] (Mar. 5, 2019). arXiv: 1802.10123. URL: http://arxiv.org/abs/1802.10123 (visited on 07/09/2020).

[WBT21]   Steen Wiewel, Moritz Becher, and Nils Thuerey. *Latent-Space Physics: Towards Learning the Temporal Evolution of Fluid Flow*. EN. https://ge.in.tum.de/publications/physics/. Feb. 2021.

[BK17]   Brunton SL, Kutz JN (2017) Data Driven Science and Engineering. 572 p. ISBN-10: 1108422098. URL: http://databookuw.com/.

Chapter 5
Physics-Informed Neural Networks

Generating an accurate surrogate model of a complex physical system usually requires a large amount of solution data about the problem at hand. However, data acquisition from experiments or simulations is often infeasible or too costly. With this in mind, Raissi et al. proposed an approach, that augments surrogate models with existing knowledge about the underlying physics of a problem [RPK19]. In many cases, the governing equations or empirically determined rules defining the problem are known a priori. For instance, an incompressible flow has to satisfy the law of conservation of mass. By incorporating this information, the solution space is drastically reduced and, as a result, less training data are needed to learn the solution.

The idea of adding prior knowledge to a machine learning algorithm is not completely new. As mentioned in Chap. 4, the studies by Raissi et al. were inspired by papers of Psichogios and Ungar [PU92], Lagaris et al. [LLF98], and more recent developments by Kondor [Kon18], Hirn et al. [HMP17] and Mallat [Mal16]. Nevertheless, the solutions proposed by Raissi et al. extended existing concepts and introduced fundamentally new approaches like a discrete time-stepping scheme, that efficiently exploits predictive power of neural networks. Furthermore, they demonstrated their method on a variety of examples that are of interest in a physics and engineering context [RPK19]. The accompanying code was written in Python and utilizes the popular GPU-accelerated machine learning framework Tensorflow. Additionally, the code is publicly available on GitHub allowing others to explore physics-informed neural networks and contribute to their development [Rai20].

The paper "Physics-informed neural networks: A deep learning framework for solving forward and inverse problems involving nonlinear partial differential equations" [RPK19] by Raissi et al. was referenced in different reviews [BNK20, FDC20]

Electronic supplementary material The online version of this chapter (https://doi.org/10.1007/978-3-030-76587-3_5) contains supplementary material, which is available to authorized users.

S. Kollmannsberger et al., *Deep Learning in Computational Mechanics*, Studies in Computational Intelligence 977, https://doi.org/10.1007/978-3-030-76587-3_5

and marks a starting point for further research on physics-enriched surrogate models
as outlined in Sect. 5.4.

5.1   Overview

The general idea of physics-informed neural networks (PINNs) is to solve problems
where only limited data are available, e.g. noisy measurements from an experiment.
To compensate for the data scarcity, the algorithm is enriched with physical laws
governing the problem at hand. Typically, those laws are described by parameterized
non-linear partial differential equations of the form

$$\frac{\partial u}{\partial t} + \mathcal{N}[u; \lambda] = 0, \ x \in \Omega, \ t \in \mathcal{T}. \tag{5.1}$$

Here, the latent solution $u(t, x)$ depends on time $t \in [0, T]$ and a spatial variable
$x \in \Omega$, where Ω refers to a space in $\mathbb{R}^D$. Further, $\mathcal{N}[u; \lambda]$ represents a non-linear
differential operator with coefficients λ. This description covers a wide variety of
problems ranging from advection–diffusion reaction of chemical or biological systems to the governing equations of continuum mechanics.

The following sections distinguish between two different use-cases for physics-
informed neural networks, namely data-driven inference and data-driven identification of partial differential equations. The first case addresses forward problems where
the coefficients λ are known and the hidden solution $u(t, x)$ is computed based on
initial and boundary data. The second case solves an inverse problem. Given scattered
data of the solution $u(t, x)$, the goal is to identify the coefficients λ of the partial
differential equation.

For stationary or transient problems, the physics-informed neural network computes the underlying partial differential equation. Therefore, a classic feed-forward
neural network approximates the solution $u(t, x)$ and the physics-enriched part evaluates the corresponding partial derivatives forming the left-hand side of Eq. (5.1).
Both the network approximating $u(t, x)$ as well as the whole physics-informed neural network depend on the same set of parameters. Those shared weights and biases
are trained by minimizing a cost function which consists of multiple mean squared
error losses. The cost function in its general form can be written as

$$C = MSE_u + MSE_f. \tag{5.2}$$

The first term, here denoted as MSE_u, computes the error of the approximation
$u(t, x)$ at known data points. In case of a forward problem, this term entails data
representing the boundary and initial conditions while for inverse problems the solution at different points inside the domain is provided. The other term in the cost
function, MSE_f, enforces the partial differential equation on a large set of randomly

chosen collocation points inside the domain. In particular, this term penalizes the error between the approximated left-hand side and the known right-hand side of the partial differential equation at every collocation point.

Generally, there are two ways to enforce boundary conditions of a forward problem. In case of a weak enforcement MSE_u, Eq. (5.2) entails multiple terms that enforce the solution at the boundary. Alternatively, regular and constant boundary conditions can be enforced in a strong sense. To this end, the output of the network is adapted so that the predicted solution automatically satisfies the boundary condition for any given input. This simplifies the cost function since the boundary terms can be neglected. As with other deep learning algorithms, the optimal network parameters are found by minimizing the custom cost function (cf. Eq. (5.2)). In the context of physics-informed neural networks gradient-based optimization algorithms like Adam or L-BFGS are often used [RPK19, Sam+19].

Even though the proposed approach is not guaranteed to converge to a global minimum, and thus, an accurate solution $u(x)$, Raissi et al. [Rai20, RPK19] showed empirically that their method achieves accurate results for different problems and architectures. Assuming that the partial differential equation has a unique solution and is well-posed, the physics-informed neural network is able to predict the underlying solution. Further requirements are a network architecture with adequate representational power and a sufficient number of collocation points N_f. In addition to the collocation-based method, Raissi et al. proposed a discrete-time model for both forward and inverse problems. This approach makes use of a discrete time-stepping scheme to predict the solution $u(x)$ at time t^{n+1} from limited snapshot data at time t^n. An advantage of the discrete method is that it omits the use of collocation points since only a solution for the initial state at t^n and boundary conditions must be provided.

The following sections explain the different types of physics-informed neural networks on forward (cf. Sect. 5.2) and backward problems (cf. Sect. 5.3). After beginning with an introductory example of a static bar, each section illustrates the continuous and discrete models with a non-linear heat problem. The implementation of the presented examples is based on the code accompanying the paper by Raissi et al. However, the code has been adapted and now utilizes the deep learning library PyTorch instead of TensorFlow.

5.2 Data-Driven Inference

In general, the problem of inference can be phrased as: find the hidden solution $u(t, x)$ for given coefficients λ. Since all values of λ are known, Eq. (5.1) simplifies to

$$\frac{\partial u}{\partial t} + \mathcal{N}[u] = 0, \quad x \in \Omega, \quad t \in \mathcal{T}. \tag{5.3}$$

5.2.1 Static Model

·A very simple application of the physics-informed neural network is the one-dimensional linear elastic static bar, which is governed by an ordinary differential equation and the corresponding boundary conditions. A solution in equilibrium state is sought so that $\frac{\partial u}{\partial t} = 0$ from Eq. (5.3). Therefore, the differential equation only applies to the spatial domain $\Omega \in \mathbb{R}^1$. The differential equation operator is then given as $\mathcal{N}[u] = \frac{d}{dx}(EA\frac{du}{dx})$. Additionally, there is an inhomogeneous term p, which defines a distributed load. The system can thereby be expressed as

$$\frac{d}{dx}\left(EA\frac{du}{dx}\right) + p = 0 \qquad\qquad \text{on } \Omega, \qquad (5.4)$$

$$EA\frac{du}{dx} = F \qquad\qquad \text{on } \Gamma_N, \qquad (5.5)$$

$$u = g \qquad\qquad \text{on } \Gamma_D. \qquad (5.6)$$

The Neumann and the Dirichlet boundaries are represented by Γ_N and Γ_D, respectively, where F denotes a concentrated load on Γ_N, and g prescribes a displacement on Γ_D. The Young's modulus $E(x)$ and the cross-sectional area $A(x)$ may vary with respect to x.

In the physics-informed neural network, a deep neural network is used to approximate the unknown solution $u(x)$. So far, this approach does not differ from the surrogates introduced in Sect. 4.3. However, these models have to rely on a tremendous amount of labeled training data closely related to the solution being approximated. Instead of having labeled training data in the whole domain, the solution for the example at hand is only known at the boundaries, for example, at $x = 0$ and $x = 1$. Additionally, Eq. (5.4) has to be satisfied for every point inside the domain. In order to verify that every input fulfills this condition, the network is extended to compute the left-hand-side of Eq. (5.4)

$$f := \frac{d}{dx}\left(EA\frac{du}{dx}\right) + p. \qquad (5.7)$$

Since a neural network is fully differentiable, it is not only possible to compute the derivatives with respect to the parameters necessary for training. Likewise, the automatic differentiation capabilities of libraries such as PyTorch or Tensorflow allow the fast computation of derivatives with respect to the input variable x (cf. Sect. 3.9).

All together, this extended network architecture can be interpreted as a physics-informed neural network with outputs $u_{NN}(x)$ and $f_{NN}(x)$. As depicted in Fig. 5.1, the first part simply approximates the solution $u(x)$ with a feed-forward fully connected neural network. The second part represents the computation of Eq. (5.7), $f(x)$ using the corresponding derivatives d_x and d_{xx} with respect to u, and the identity I of u. It should be noted that both networks $u_{NN}(x)$ and $f_{NN}(x)$ depend on the same set of parameters, namely the weights and biases of the first network. The dashed

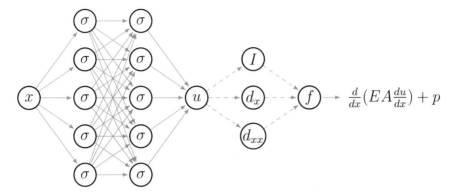

Fig. 5.1 Conceptual physics-informed neural network for the static bar equation. The left part shows the feed-forward neural network and the right part represents the physics-informed neural network. The dashed lines denote non-trainable weights

connections in Fig. 5.1 are simply introduced to visualize the composition of f and do not represent trainable parameters of the network.

To give a better idea of how a physics-informed neural network is implemented, a specific example is now introduced. Let us define a domain $\Omega = [0, 1]$ with $\Gamma_D = \{x \mid x = 0, \ x = 1\}$ and $\Gamma_N = \emptyset$. The material parameters are set to $EA = 1$. A solution for the displacement is chosen as

$$u(x) = \sin(2\pi x), \tag{5.8}$$

which, after insertion into the differential equation (5.4), results in the distributed load

$$p(x) = 4\pi^2 \sin(2\pi x). \tag{5.9}$$

Furthermore, the following Dirichlet boundary conditions apply

$$u(0) = u(1) = 0. \tag{5.10}$$

With this information it is now possible to define a cost function. In case of the static physics-informed neural network this custom cost function is assembled from two mean squared error losses Eq. (5.2)

$$C = MSE_b + MSE_f, \tag{5.11}$$

where

$$MSE_b = \frac{1}{N_b} \sum_{i=1}^{N_b} \left(u_{NN} \left(x_b^i \right) - u_b^i \right)^2, \tag{5.12}$$

and

$$MSE_f = \frac{1}{N_f} \sum_{i=1}^{N_f} \left(f_{NN} \left(x_f^i \right) \right)^2 . \qquad (5.13)$$

The solution on the boundary is represented by N_b labeled data points $\{x_b^i, u_b^i\}_{i=1}^{N_b}$. The boundary loss MSE_b is computed by comparing the approximation u_{NN} with the labels u_b of the training data. The error of the approximation at known data points is given by $MSE_u = MSE_b$, as only the boundary data points are known. Note also, that here the boundary loss MSE_b solely includes Dirichlet boundary conditions, as only these were defined in the example. However, the application of Neumann boundary conditions works similarly and is discussed further in Sect. 5.2.2.

In order to enforce Eq. (5.3) in the whole spatial domain, a set of N_f colloca-tion points $\{x_f^i\}_{i=1}^{N_f}$ is generated using a uniform distribution throughout the one-dimensional domain. The corresponding loss MSE_f from Eq. (5.13) is computed as the mean squared error of f_{NN} at all collocation points. Now, the physics-informed neural network can be trained by minimizing the cost function in Eq. (5.11). Instead of a classical stochastic gradient descent procedure, the L-BFGS optimizer is employed [LN89].

The physics-informed network is implemented in PyTorch. At first, a neural net-work is created to predict the displacement $u(x)$. For simplicity's sake, a model consisting of a single hidden layer with the dimension `hidden_dim` is defined. The input and output dimensions are correspondingly given as `input_dim` and `output_dim`. The linear transformations from one layer to the next are defined by `torch.nn.Linear` and the non-linear activation function is chosen as the hyper-bolic tangent, `torch.nn.Tanh`. There is no activation function in the output layer meaning that the range of the outputs is not limited.

```
def buildModel(input_dim, hidden_dim, output_dim):
    model = torch.nn.Sequential(torch.nn.Linear(input_dim, hidden_dim),
    torch.nn.Tanh(),
    torch.nn.Linear(hidden_dim, output_dim))
    return model
```

With the neural network, the first part of the PINN illustrated in Fig. 5.1 is defined. A displacement prediction $\hat{u}$ at $x = 0.5$ can be made with the following code.

```
model = buildModel(1, 10, 1)
u_pred = model(torch.tensor([0.5]))
```

The second part of the network requires the computation of the derivatives with respect to the input x. This is achieved with PyTorch's built-in automatic differ-entiation, which creates and retains a computational graph to store all information necessary for calculating the gradients.

```
def get_derivative(y, x):
    dydx = grad(y, x, torch.ones(x.size()[0], 1),
        create_graph=True,
        retain_graph=True)[0]
    return dydx
```

Finally, $f(x)$ from Eq. (5.7) is computed with the following code.

```
def f(model, x, EA, p):
    u = model(x)
    u_x = get_derivative(u, x)
    EAu_xx = get_derivative(EA(x) * u_x, x)
    f = EAu_xx + p(x)
    return f
```

Now, to calculate the loss function of the differential equation MSE_f, the example values are inserted into the model.

```
model = buildModel(1, 10, 1)
x = torch.linspace(0, 1, 10, requires_grad=True).view(-1, 1)
EA = lambda x: 1 + 0 * x
p = lambda x: 4 * math.pi**2 * torch.sin(2 * math.pi * x)

f = f(model, x, EA, p)
MSE_f = torch.sum(f**2)
```

In combination with the loss function of the boundary conditions MSE_b, allowing to compute the cost function C from Eq. (5.11). The implementation for this example is shown in the code snippet below. Note that only the Dirichlet boundary conditions are imposed here. Neumann boundary conditions can be imposed in a similar way with the small addition of computing the derivatives with the `get_derivative` function. This is described in more detail in the upcoming Sect. 5.2.2.

```
model = buildModel(1, 10, 1)
u0 = 0
u1 = 0

u0_pred = model(torch.tensor([0.]))
u1_pred = model(torch.tensor([1.]))
MSE_b = (u0_pred - u0)**2 + (u1_pred - u1)**2
```

It is now possible to compute and minimize the cost function C via a stochastic gradient approach or more sophisticated methods such as the L-BFGS optimizer. When the cost function is sufficiently small, an accurate prediction of the displacement

$u(x)$ is to be expected. In principle, this is how the physics informed neural network works. The entire training procedure is summarized in Algorithm 4.

Algorithm 4 Training a physics-informed neural network for the static solution of the problem described in Eq. (5.4).

Require: training data for boundary condition $\{x_b^i, u_b^i\}_{i=1}^{N_b}$

 generate N_f collocation points with a uniform distribution $\{x_f^i\}_{i=1}^{N_f}$
 define network architecture (input, output, hidden layers, hidden neurons)
 initialize network parameters $\boldsymbol{\Theta}$: weights $\{\boldsymbol{W}^l\}_{l=1}^{L}$ and biases $\{\boldsymbol{b}^l\}_{l=1}^{L}$ for all layers L
 set hyperparameters for L-BFGS optimizer (*epochs*, learning rate α, …)

 for all *epochs* **do**
   $\hat{\boldsymbol{u}}_b \leftarrow \boldsymbol{u}_{NN}(\boldsymbol{x}_b; \boldsymbol{\Theta})$
   $\boldsymbol{f} \leftarrow \boldsymbol{f}_{NN}(\boldsymbol{x}_f; \boldsymbol{\Theta})$
   compute MSE_b, MSE_f ▷ cf. Eqs. (5.12) and (5.13)
   compute cost function: $C \leftarrow MSE_b + MSE_f$
   update parameters: $\boldsymbol{\Theta} \leftarrow \boldsymbol{\Theta} - \alpha \frac{\partial C}{\partial \boldsymbol{\Theta}}$ ▷ L-BFGS
 end for

Alternatively, it is possible to use a strong enforcement of the Dirichlet boundary conditions to simplify the minimization problem. For that, the output of the neural network must be adapted to automatically satisfy the boundary conditions for any given input. Then, the mean squared error loss of the Dirichlet boundary conditions can be dropped. The strong enforcement of boundary conditions is demonstrated on a transient heat problem in Sect. 5.2.3.

The static bar example is now used to build up a PINN. The corresponding results are presented in Fig. 5.2. Here, the boundary conditions are enforced via the cost function. The plot on the left shows that the estimated displacements and the analytic displacements coincide. Additionally, the training history is illustrated on the right, showing the cost function for every training epoch. One observes that there is a difference in the order of magnitudes between the cost function of the differential equation and the cost function of the boundary conditions. Adjusting this difference with a weighting factor on the boundary condition term is possible, yet difficult to determine a priori. An alternative would be to enforce the Dirichlet boundary conditions strongly, which in this case leads to an unconstrained optimization problem, as only the differential equation cost has to be minimized.

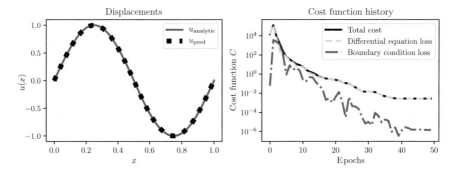

Fig. 5.2 Linear elastic static bar example. Left: the displacements $u(x)$ computed with a physics-informed neural network and the analytic solution are compared. Right: the training history is illustrated by showing the cost function for each epoch

5.2.2 Continuous-Time Model

To demonstrate the application of PINNs to transient problems, an initial-boundary value problem described by a one-dimensional non-linear heat equation is introduced. The evolution of temperature $u(t, x)$ as a function of time t and space x is described by a partial differential equation and boundary conditions of the following form

$$c\frac{\partial u}{\partial t} - \frac{\partial}{\partial x}\left(\kappa\frac{\partial u}{\partial x}\right) - s = 0 \qquad \text{on } \mathcal{T} \times \Omega, \qquad (5.14)$$

$$\kappa\frac{\partial u}{\partial x} = h \qquad \text{on } \mathcal{T} \times \Gamma_N, \qquad (5.15)$$

$$u = g \qquad \text{on } \mathcal{T} \times \Gamma_D, \qquad (5.16)$$

$$u(x, 0) = u_0 \qquad \text{on } \Omega. \qquad (5.17)$$

Here, $\Omega \subset \mathbb{R}^1$ represents the one-dimensional spatial domain, whereas $\mathcal{T}$ denotes the temporal domain. Together they form the time-space domain $\mathcal{T} \times \Omega$ with Neumann and Dirichlet boundaries represented by Γ_N and Γ_D, respectively. In engineering applications the heat capacity $c(u)$ or the thermal conductivity $\kappa(u)$ are often temperature-dependent which introduces a non-linearity to Eq. (5.3).

The specific problem investigated in the following is defined on the time-space domain $\mathcal{T} \times \Omega = t \in [0, 0.5] \times x \in [0, 1]$ with $\Gamma_D = \emptyset$ and $\Gamma_N = \{x \mid x = 0, x = 1\}$. In particular, the problem is subject to homogeneous Neumann boundary conditions

$$\frac{\partial u(t, 0)}{\partial x} = \frac{\partial u(t, 1)}{\partial x} = 0, \qquad (5.18)$$

and the initial condition

$$u(0, x) = u_0. \qquad (5.19)$$

Both the heat capacity $c(u)$ and the thermal conductivity $\kappa(u)$ are temperature-dependent and defined as [Kol+18]

$$c(u) = 1/2000 \, u^2 + 500, \tag{5.20}$$

$$\kappa(u) = 1/100 \, u + 7. \tag{5.21}$$

In order to verify the network predictions, a manufactured solution of the following form is proposed

$$u = \exp\left(-\frac{(x - p)^2}{2\sigma^2}\right), \tag{5.22}$$

where $\sigma = 0.02$. Normally, such Gaussian bell formulations are used to model a laser-induced heat source s [Kol+19] which is traveling along a path p

$$p(t) = \frac{1}{4} \cos\left(\frac{2\pi t}{t_{\max}}\right) + \frac{1}{2}. \tag{5.23}$$

Here, the Gaussian bell term is chosen as the manufactured solution to the problem instead. Plugging Eq. (5.22) into (5.14) yields

$$s = \frac{\kappa u}{\sigma^2} + u \frac{x - p}{\sigma^2} \left[c \frac{\partial p}{\partial t} - \frac{x - p}{\sigma^2}\left(\kappa + u \frac{\partial \kappa}{\partial u}\right)\right]. \tag{5.24}$$

The method of manufactured solutions is commonly used to verify numerical solvers on sufficiently complex examples [Roa02]. Nonetheless, the manufactured solution from Eq. (5.22) should by no means be interpreted as an realistic example. It simply serves as a fast and direct way to benchmark the predictions of the physics-informed neural network. For the example at hand the benchmark solution u and the corresponding heat flux ϕ were generated with MATLAB at 201×256 discrete points in the time-space domain (cf. Fig. 5.3).

The implementation that generated the following results is based on the code accompanying the paper by Raissi et al. [Rai20, RPK19], but was adapted to employ the PyTorch framework. Like in the static example, a feed-forward neural network $u_{NN}(t, x; \boldsymbol{\Theta})$ is used to predict the hidden solution of the problem specified in Eqs. (5.14), (5.19), (5.18). The inputs of the network are the temporal variable t and the spatial variable x. The function `build_model` (cf. Sect. 5.2.1) determines the layer and neuron architecture of the feed-forward neural network u_{NN} which approximates the temperature distribution $u(t, x)$. To define the output of the physics-informed neural network $f_{NN}(t, x; \boldsymbol{\Theta})$ (cf. Fig. 5.4), the left-hand side of Eq. (5.14) is recalled as

$$f := c \frac{\partial u}{\partial t} - \frac{\partial}{\partial x}\left(\kappa \frac{\partial u}{\partial x}\right) - s = c \frac{\partial u}{\partial t} - \frac{\partial \kappa}{\partial u} \frac{\partial u}{\partial x} \frac{\partial u}{\partial x} - \kappa \frac{\partial^2 u}{\partial x^2} - s. \tag{5.25}$$

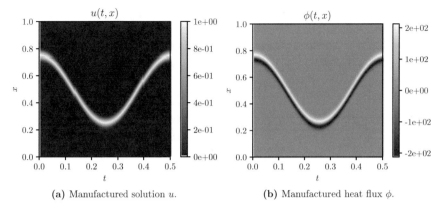

(a) Manufactured solution u.   (b) Manufactured heat flux ϕ.

Fig. 5.3 Manufactured solution u and corresponding flux ϕ for the one-dimensional heat transfer problem. The solution was generated using MATLAB with a resolution of $t \times x = 201 \times 256$ points

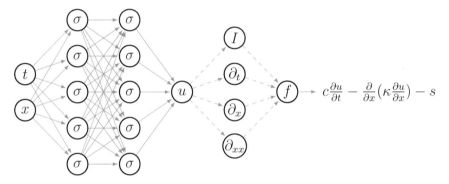

Fig. 5.4 Conceptual physics-informed neural network for the heat equation. The left side shows the feed-forward neural network and the right side represents the physics-informed part. The dashed lines denote non-trainable weights

The code corresponding to Eq. (5.25) is given as

```
def f_nn(self, t, x):
    u = self.u_nn(t, x)
    u_t = get_derivative(u, t, 1)
    u_x = get_derivative(u, x, 1)
    u_xx = get_derivative(u, x, 2)
    k = 0.01 * u + 7
    k_u = 0.01
    c = 0.0005 * u ** 2 + 500
    s = self.source_term(t,x)
    f = c * u_t - k_u * u_x * u_x - k * u_xx - s
    return f
```

Here, `self.u_nn(t,x)` calls the neural network $u_{NN}(t, x; \boldsymbol{\Theta})$ to return the predicted temperature u. Then, the previously introduced function `get_derivative` (cf. Sect. 5.2.1) is extended to recursively compute the necessary first- and second-order partial derivatives.

```
def get_derivative(y, x, n):
    if n == 0:
        return y
    else:
        dy_dx = grad(y, x, torch.ones_like(y), create_graph=True,
                     retain_graph=True, allow_unused=True)[0]
        return get_derivative(dy_dx, x, n - 1)
```

Furthermore, `self.source_term(t,x)` returns the source term s according to Eq. (5.24).

```
def source_term(self, t, x):
    t_max = 0.5
    sigma = 0.02
    u_max = 1
    p = 0.25 * torch.cos(2 * np.pi * t / t_max) + 0.5
    p_t = -0.5 * torch.sin(2 * np.pi * t / t_max) * np.pi / t_max
    u_sol = u_max * torch.exp(-(x - p) ** 2 / (2 * sigma ** 2))
    k_sol = 0.01 * u_sol + 7
    k_u_sol = 0.01
    c_sol = 0.0005 * u_sol ** 2 + 500
    factor = 1/(sigma ** 2)
    s = factor * k_sol * u_sol
        + u_sol * (x - p) * factor * (c_sol * p_t
        - (x - p) * factor * (k_sol + u_sol * k_u_sol))
    return s
```

In order to train the network, the following cost function is defined

$$C = MSE_0 + MSE_b + MSE_f. \tag{5.26}$$

Here, the term MSE_0

$$MSE_0 = \frac{1}{N_0} \sum_{i=1}^{N_0} \left(u_{NN}(0, x_0^i; \boldsymbol{\Theta}) - u_0^i \right)^2, \tag{5.27}$$

enforces the initial condition (cf. Eq. (5.19)) by penalizing the error between the network prediction $\{u_{NN}(0, x_0^i, \boldsymbol{\Theta})\}_{i=1}^{N_0}$ and the initial solution $\{u_0^i\}_{i=1}^{N_0}$ at N_0 points randomly drawn from a uniform distribution. The term MSE_b

$$MSE_b = \frac{1}{N_b} \sum_{i=1}^{N_b} \left(\frac{\partial}{\partial x} u_{NN}\left(t_b^i, 0; \boldsymbol{\Theta}\right) \right)^2 + \frac{1}{N_b} \sum_{i=1}^{N_b} \left(\frac{\partial}{\partial x} u_{NN}\left(t_b^i, 1; \boldsymbol{\Theta}\right) \right)^2, \tag{5.28}$$

enforces the Neumann boundary condition according to Eq. (5.18) at N_b random samples $\{t_b, x_b^i\}_{i=1}^{N_b}$ on each boundary $x = 0$ and $x = 1$. Finally, adding the error of the residual MSE_f

$$MSE_f = \frac{1}{N_f} \sum_{i=1}^{N_f} \left(f_{NN} \left(t_f^i, x_f^i \right) \right)^2 \tag{5.29}$$

ensures that the solution satisfies the governing Eq. (5.14). As proposed by Raissi et al. [RPK19], the N_f collocation points $\{t_f, x_f^i\}_{i=1}^{N_f}$ are generated by a Latin-hypercube sampling technique [Ste87]. Following Eqs. (5.27)–(5.29) the cost function is implemented as follows.

```
def cost_function(self, t0, x0, t_lb, x_lb, t_ub, x_ub, t_f, x_f, u0):
    u0_pred = self.u_nn(t0, x0)
    u_lb_pred = self.u_nn(t_lb, x_lb)
    u_x_lb_pred = get_derivative(u_lb_pred, x_lb, 1)
    u_ub_pred = self.u_nn(t_ub, x_ub)
    u_x_ub_pred = get_derivative(u_ub_pred, x_ub, 1)
    f_pred = self.f_nn(t_f, x_f)
    mse_0 = torch.mean((u0 - u0_pred)**2)
    mse_b = torch.mean(u_x_lb_pred**2) + torch.mean(u_x_ub_pred**2)
    mse_f = torch.mean((f_pred)**2)
    return mse_0, mse_b, mse_f
```

Subsequently, a full-batch gradient-based optimization procedure searches for the optimal network parameters Θ^* by minimizing the cost function in Eq. (5.26). In addition to the L-BFGS method, a preceding minimization with the Adam optimizer is employed [KB17]. This combination was adapted from [Rai20] and has empirically proven to be the most robust approach throughout this study.

Algorithm 5 summarizes the previously introduced training procedure employed for a continuous prediction of the temperature distribution $u(t, x)$. Figure 5.5a shows the history of the cost function terms. After the algorithm terminated, the network $u_{NN}(t, x; \Theta)$ with trained parameters Θ is used to predict the continuous temperature distribution over the whole domain $\mathcal{T} \times \Omega = [0, 0.5] \times [0, 1]$. The result along with a comparison between the prediction and the manufactured solution is presented in Fig. 5.6.

When optimizing a cost function that consists of multiple terms, it is important to balance the influence of each term. Figure 5.5b shows an example where no balancing is applied. The term MSE_f is magnitudes larger than the two other terms MSE_0 and MSE_b. Here, the initial and boundary conditions are underrepresented in the cost function and the algorithm does not converge to a correct solution (cf. Fig. B.1). Nabian and Meidani addressed this issue by introducing weight factors adjusting the relative importance of each term in the cost function [NM19]. In the example at hand, an empirically determined factor of 5×10^{-4} must be added to the term mse_f in cost_function to gain satisfactory results (cf. Figs. 5.5a and 5.6).

```
mse_f = torch.mean((5e-4*f_pred)**2)
```

Algorithm 5 Training a physics-informed neural network for the continuous solution of the problem described in Eq. (5.14).

Require: training data for initial condition $\{0, x_0^i, u_0^i\}_{i=1}^{N_0}$
Require: training data for boundary condition $\{t_b, x_b^i, u_b^i\}_{i=1}^{N_b}$

  generate N_f collocation points with Latin-hypercube sampling $\{t_f, x_f^i\}_{i=1}^{N_f}$
  define network architecture (input, output, hidden layers, hidden neurons)
  initialize network parameters $\boldsymbol{\Theta}$: weights $\{W^l\}_{l=1}^L$ and biases $\{b^l\}_{l=1}^L$ for all layers L
  set hyperparameters for Adam optimizer (Adam-*epochs*, learning rate α, ...)
  set hyperparameters for L-BFGS optimizer (L-BFGS-*epochs*, convergence criterion, ...)

  procedure TRAIN
    compute $\{u_{NN}(0, x_0^i, \boldsymbol{\Theta})\}_{i=1}^{N_0}$
    compute $\{\frac{\partial}{\partial x} u_{NN}(t_b^i, x_b^i; \boldsymbol{\Theta})\}_{i=1}^{N_b}$
    compute $\{f_{NN}(t_f^i, x_f^i; \boldsymbol{\Theta})\}_{i=1}^{N_f}$
    compute MSE_0, MSE_b, MSE_f ▷ cf. Eqs. (5.27) to (5.29)
    evaluate cost function: $C \leftarrow MSE_0 + MSE_b + MSE_f$
    update parameters: $\boldsymbol{\Theta} \leftarrow \boldsymbol{\Theta} - \alpha \frac{\partial C}{\partial \boldsymbol{\Theta}}$ ▷ Adam or L-BFGS
  end procedure

  for all Adam-*epochs* **do**
    run TRAIN with Adam optimizer
  end for
  for all L-BFGS-*epochs* **do**
    run TRAIN with L-BFGS optimizer
  end for

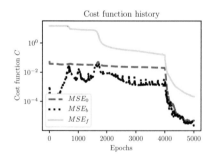

(a) Balanced cost function terms plotted over the number of training iterations.

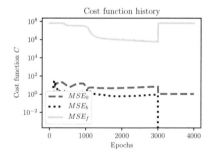

(b) Unbalanced cost function terms plotted over the number of training iterations.

Fig. 5.5 Balanced and unbalanced cost function history for 3000 Adam epochs and 1000 L-BFGS epochs

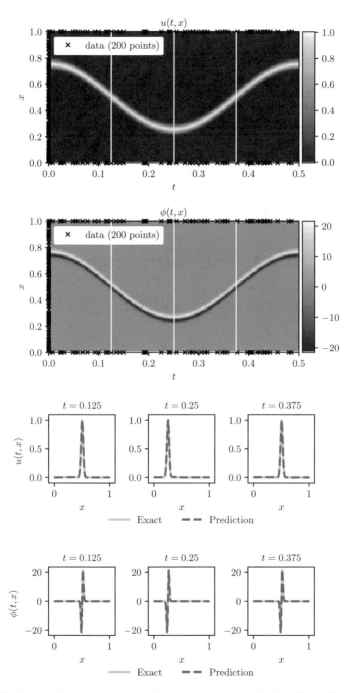

Fig. 5.6 Prediction of the temperature distribution u and corresponding heat flux ϕ. Top: approximated solution and location of time snapshots (white lines). Bottom: comparison of predicted and exact solution at distinct snapshots

5.2.3 Discrete-Time Model

To circumvent the need for collocation points Raissi et al. proposed an alternative solution-inference approach based on a Runge–Kutta time-stepping scheme. Contrary to a continuous prediction in time, the solution is only predicted at specific time steps. Assuming the solution at time step t^n is known, then the proposed method is able to predict the solution at the next time step $t^{n+1} = t^n + \Delta t$, where Δt denotes the step size.

To solve the problem

$$\frac{\partial u}{\partial t} = g[u], \tag{5.30}$$

the general form of the Runge–Kutta method with q stages is [Ise08]

$$u^{n+c_i} = u^n + \Delta t \sum_{j=1}^{q} a_{ij} g \left[u^{n+c_j} \right], \quad i = 1, \dots, q,$$

$$u^{n+1} = u^n + \Delta t \sum_{j=1}^{q} b_j g \left[u^{n+c_j} \right], \tag{5.31}$$

where

$$u^{n+c_j}(x) = u(t^n + c_j \Delta t, x) \quad j = 1, \dots, q. \tag{5.32}$$

Writing the Runge–Kutta method in this general form allows the usage of both explicit and implicit time-stepping schemes. The type of method depends on the choice of parameters a_{ij}, b_j, and c_j that are organized in a so-called Butcher table [Ise08]. The advantage of explicit methods is that they are fast and easy to implement. After the solution at the first stage is obtained, it is substituted into the equation at the second stage and so on. As the name suggests, implicit methods can not simply be solved by substitution. They form a system of dependent equations that requires the use of an iterative solution process. However, implicit methods exhibit excellent stability properties, which make them especially suitable for stiff systems.

Assuming a feed-forward neural network u_{NN}^{n+1} is able to predict the solution u^{n+1} at time t^{n+1} and the intermediate solutions u^{n+c_i} at all stages $i = 1, \dots, q$ from an input x, then its output can be written as

$$\left[u^{n+c_1}(x), \dots, u^{n+c_q}(x), u^{n+1}(x) \right] \leftarrow u_{NN}^{n+1}. \tag{5.33}$$

In particular, the neural network predicts the left-hand side of Eq. (5.31). Rearranging Eq. (5.31) yields

$$u^n = u^{n+c_i} - \Delta t \sum_{j=1}^{q} a_{ij} g \left[u^{n+c_j} \right], \quad i = 1, \ldots, q,$$

$$u^n = u^{n+1} - \Delta t \sum_{j=1}^{q} b_j g \left[u^{n+c_j} \right]. \tag{5.34}$$

Now, all the terms dependent on the prediction of the neural network stand on the right-hand side and the solution at time t^n is found on the left. In other words, the time-stepping scheme is reversed to get an estimate of u^n that is dependent on the neural network prediction u_{NN}^{n+1}. The error between this estimate and the known solution at time t^n is later used to formulate a cost function for training the network parameters. To assign a unique identifier to each equation in Eq. (5.34), the following nomenclature is introduced

$$u^n = u_i^n, \quad i = 1, \ldots, q,$$

$$u^n = u_{q+1}^n, \tag{5.35}$$

where

$$u_i^n = u^{n+c_i} - \Delta t \sum_{j=1}^{q} a_{ij} g \left[u^{n+c_j} \right], \quad i = 1, \ldots, q$$

$$u_{q+1}^n = u^{n+1} - \Delta t \sum_{j=1}^{q} b_j g \left[u^{n+c_j} \right]. \tag{5.36}$$

Eventually, Eqs. (5.35) and (5.36) define the output of the physics-informed neural network u_{NN}^n for an input x as follows

$$\left[u_1^n(x), \ldots, u_q^n(x), u_{q+1}^n(x) \right] \leftarrow u_{NN}^n. \tag{5.37}$$

Like in the continuous-time model, the physics-informed neural network consists of two parts (cf. Fig. 5.7). The first part is a deep neural network with multiple outputs as shown in (5.33). The second part transforms the output of the first network according to Eqs. (5.35) and (5.36) and returns the quantities in (5.37) for comparison with the known solution u^n at time t^n.

All in all, the network learns to predict the solution at time t^{n+1} based on the known solution at time t^n and the boundary conditions in the time interval $[t^n, t^{n+1}]$ by minimizing a suitable cost function. This step can be repeated to predict the solution at the following time steps $u(t^{n+2}, x)$, $u(t^{n+3}, x)$, and so on. When explicit methods are used, the step size Δt is chosen to be small in order to prevent stability issues. Implicit schemes are stable even for larger time steps. In classic, implicit discretization schemes, this simultaneously leads to a costly increase in the number of required stages q. However, what makes the proposed method distinct from the classical Runge–Kutta time-stepping schemes is the fact that the number of stages q can be increased without a significant increase in computational effort. Adding

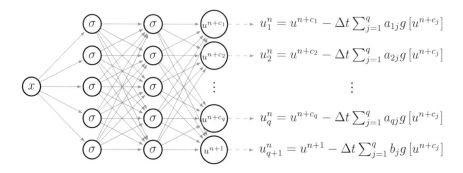

Fig. 5.7 Discrete physics-informed neural network with one input and $q + 1$ outputs. The feed-forward neural network generates a prediction of u^{n+1} and the intermediate solutions u^{n+c_i}, $i = 1, \ldots, q$ for q stages, which are then used in Eqs. (5.35) and (5.36) to compute an output that can be compared to the solution u^n at initial time t^n

another stage to the model simply extends the output layer of the neural network by an extra neuron and the corresponding parameters. Overall, the parameters only increase linearly with the total number of stages.

To study the discrete solution method, the heat problem from the previous Sect. 5.2.2 is revisited. The governing partial differential equation is observed in the time-space domain $\mathcal{T} \times \Omega = [0, 0.5] \times [0, 1]$ with $\Gamma_D = \{x \mid x = 0, \ x = 1\}$ and $\Gamma_N = \emptyset$ and is written as

$$c \frac{\partial u}{\partial t} - \frac{\partial}{\partial x} \left(\kappa \frac{\partial u}{\partial x} \right) - s = 0 \quad t \in [0, 0.5], \ x \in [0, 1]. \tag{5.38}$$

The temperature-dependent coefficients, namely, the heat capacity $c(u)$ and thermal conductivity $\kappa(u)$ are defined as [Kol+18]

$$c(u) = 1/2000 \, u^2 + 500, \tag{5.39}$$
$$\kappa(u) = 1/100 \, u + 7. \tag{5.40}$$

In contrast to the case studied in Sect. 5.2.2, the problem is now subject to homogeneous Dirichlet boundary conditions

$$u(0, t) = u(1, t) = 0. \tag{5.41}$$

Further, the problem is subject to the following initial condition

$$u(0, x) = u_0. \tag{5.42}$$

To validate the computation and to provide an initial solution snapshot at time t^n, the manufactured solution (cf. Eq. (5.22)) and the corresponding source term (cf. Eq. (5.24)) from Sect. 5.2.2 are reused. The implementation of the discrete method is inspired by the code from Raissi [Rai20], but uses the PyTorch library to construct the physics-informed neural network.

At first, the architecture of the physics-informed surrogate model for a desired time step $t^{n+1} = t^n + \Delta t$ and q stages is specified. Since a discretization in time is applied, the feed-forward neural network $u_{NN}^{n+1}(x; \Theta)$ only takes the spatial variable x as an input. Next to the desired solution $u^{n+1}(x)$ at time t^{n+1}, the network predicts the intermediate solutions u^{n+c_i} for all stages $i = 1, \dots, q$. Here, the parameters a_{ij}, b_j, and c_j for the implicit Runge–Kutta time-stepping scheme are taken from a Butcher table corresponding to the number of stages [Ise08]. To apply the temporal discretization, the problem described in Eq. (5.38) is rearranged in accordance with Eq. (5.30)

$$\frac{\partial u}{\partial t} = \frac{1}{c}\left(\frac{\partial \kappa}{\partial u}\frac{\partial u}{\partial x}\frac{\partial u}{\partial x} + \kappa\frac{\partial^2 u}{\partial x^2} + s\right). \tag{5.43}$$

Thus, $g\left[u^{n+c_j}\right]$ can be defined as

$$g\left[u^{n+c_j}\right] = \frac{1}{c}\left(\frac{\partial \kappa}{\partial u^{n+c_j}}\frac{\partial u^{n+c_j}}{\partial x}\frac{\partial u^{n+c_j}}{\partial x} + \kappa\frac{\partial^2 u^{n+c_j}}{\partial x^2} + s\right), \tag{5.44}$$

where

$$c(u) = 1/2000\,(u^{n+c_j})^2 + 500, \tag{5.45}$$

$$\kappa(u) = 1/100\,u^{n+c_j} + 7. \tag{5.46}$$

Following the modified Runge–Kutta time-stepping scheme from Eqs. (5.35) and (5.36) and using Eq. (5.44), the physics-informed neural network $u_{NN}^n(x; \Theta)$ with outputs $\left[\hat{u}_1^n(x), \dots, \hat{u}_q^n(x), \hat{u}_{q+1}^n(x)\right]$ is defined.

```
def U0_nn(self, x):
    U1 = self.U1_nn(x)
    U = U1[:, :-1]
    U_x = torch.zeros_like(U)
    U_xx = torch.zeros_like(U)
    for i in range(U.size(1)):
        U_x[:,i:i+1] = get_derivative(U[:,i], x, 1)
        U_xx[:,i:i+1] = get_derivative(U[:,i], x, 2)

    t = self.t0 + self.dt * self.IRK_times.T
    s = self.source_term(t, x)

    c = 0.0005 * U ** 2 + 500
```

```
k = 0.01 * U + 7
k_u = 0.01
F = (k_u * U_x * U_x + k * U_xx + s) / c
U0 = U1 - self.dt * torch.matmul(F, self.IRK_weights.T)
return U0
```

Here, the feed-forward neural network `self.U1_nn` predicts the solution at time t^{n+1}. Its implementation is explained in more detail below. The variable `self.t0` stores the initial time t^n while `self.dt` corresponds to the time step size Δt. The parameters for the implicit Runge–Kutta time-stepping scheme a_{ij}, b_j are arranged in the matrix `self.IRK_weights`, while the vector `self.IRK_times` contains the parameters c_j.

So far, the initial and boundary conditions have been enforced weakly in the cost function. However, it is also possible to apply constraints in a strong sense. To do so, the solution u is modified to satisfy the boundary conditions for any given input. Following the generalized approach of Lagaris et al. [LLF98] for a strong enforcement, the network output is multiplied by $(1 - x)x$ to consider for the homogeneous Dirichlet boundary conditions as prescribed in Eq. (5.41). Since the discrete solution is only dependent on the spatial variable x, the output can be written as

$$\left[\hat{u}^{n+c_1}, \ldots, \hat{u}^{n+c_q}, \hat{u}^{n+1}(x)\right] \leftarrow (1 - x)x\, \boldsymbol{u}_{NN}^{n+1}(x; \boldsymbol{\Theta}) \quad i = 1, \ldots, q, \qquad (5.47)$$

which easily translates into the following code.

```
def U1_nn(self, x):
    U1 = (1-x)*x*self.model(x)
    return U1  # N x (q+1)
```

Due to the discretization in time, no collocation points are needed to train the physics-informed neural network. Hence, the term MSE_f from Eq. (5.2) can be dropped. Moreover, the introduction of a strong boundary enforcement obviates the boundary term MSE_b as part of MSE_u. As a result, the cost function for training the network simplifies to a single loss term eliminating the need to determine suitable weighting factors. With the output of the physics-informed neural network $\left[\hat{u}_1^n(x), \ldots, \hat{u}_q^n(x), \hat{u}_{q+1}^n(x)\right] \leftarrow \boldsymbol{u}_{NN}^n(x; \boldsymbol{\Theta})$ (cf. (5.37)), the resulting cost function is written as

$$C = MSE_n, \qquad (5.48)$$

where

$$MSE_n = \sum_{j=1}^{q+1} \sum_{i=1}^{N_n} \left(\hat{u}_j^n\left(x^i\right) - u^{n,i}\right)^2. \qquad (5.49)$$

The term MSE_n computes the prediction error of the physics-informed neural network at N_n randomly sampled points $\{x^{n,i}, u^{n,i}\}_{i=1}^{N_n}$ of the solution at initial time t^n.

This could be the initial condition of the problem at hand or any other snapshot of the solution, e.g. at time $t^n = 0.05$. According to Eq. (5.49), the cost function can be implemented as follows:

```python
def cost_function(self, x0, u0):
    U0_pred = self.U0_nn(x0)
    return torch.mean((U0_pred - u0)**2)
```

To learn the shared set of optimal parameters Θ^*, the combination of Adam optimizer and L-BFGS method is employed to minimize the cost function introduced in Eq. (5.48).

Finally, Algorithm 6 summarizes the previously elaborated steps for training the discrete-time physics-informed neural network. After the algorithm terminated, the network $u_{NN}^{n+1}(x; \Theta)$ with the trained parameters Θ is used to predict the temperature at time t^{n+1} for a given input x. For a prediction of the temperature distribution $u(x)$ at time $t^{n+1} = 0.3$ given a snapshot of $N_n = 200$ random data points at time $t^n = 0.05$, the cost function history and the results are shown in Figs. 5.8 and 5.9, respectively.

Algorithm 6 Training a discrete physics-informed neural network for a single time step $t^{n+1} = t^n + \Delta t$ and q stages.

Require: N_n training samples for initial snapshot $\{x^i, u^{n,i}\}_{i=1}^{N_n}$ at time t^n
    define time step size Δt
    define number of stages q
    define network architecture (input, hidden layers, hidden neurons)
    initialize output layer with $q + 1$ neurons
    initialize network parameters Θ: weights $\{W^l\}_{l=1}^{L}$ and biases $\{b^l\}_{l=1}^{L}$ for all layers L
    set hyperparameters for Adam optimizer (Adam-*epochs*, learning rate α, ...)
    set hyperparameters for L-BFGS optimizer (L-BFGS-*epochs*, convergence criterion, ...)

    procedure TRAIN
        compute $\left\{\left[\hat{u}_1^n(x^i), \ldots, \hat{u}_q^n(x^i), \hat{u}_{q+1}^n(x^i)\right] \leftarrow u_{NN}^n(x^i; \Theta)\right\}_{i=1}^{N_n}$
        compute MSE_n ▷ cf. Eq. (5.49)
        $C \leftarrow MSE_n$
        update parameters: $\Theta \leftarrow \Theta - \alpha \frac{\partial C}{\partial \Theta}$ ▷ Adam or L-BFGS
    end procedure
    for all Adam-*epochs* **do**
        run TRAIN with Adam optimizer
    end for
    for all L-BFGS-*epochs* **do**
        run TRAIN with L-BFGS optimizer
    end for

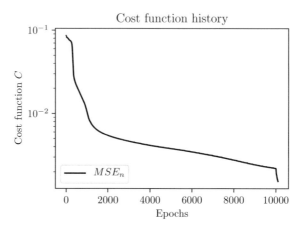

Fig. 5.8 Cost function history of discrete-time inference physics-informed neural network

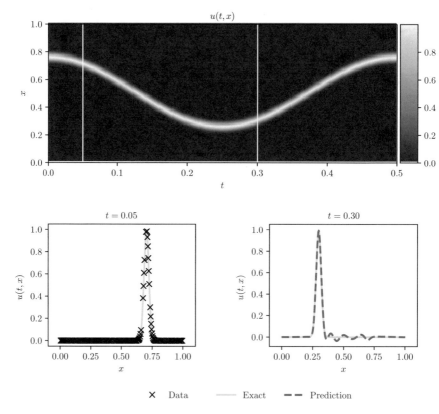

Fig. 5.9 Predictions of the inference physics-informed neural network. Top: manufactured solution u and time snapshots $t^n = 0.05$ and $t^{n+1} = 0.3$ (white lines). Bottom: training data at $t^n = 0.05$ and prediction at $t^{n+1} = 0.3$

5.3 Data-Driven Identification

System identification from sparse data is another class of problems often encountered in physics or engineering applications. In the second part of their article, Raissi et al. addressed the problem of data-driven discovery with the help of the previously introduced physics-informed neural networks. In other words, the task of system identification can be phrased as: find the parameters λ that describe the observed data best. The main idea is to use the non-linear partial differential equations (5.1) together with a solution $u(t, x)$ to solve the inverse problem of identifying the parameters λ.

A neural network is used to predict the parameter λ instead of the solution $u(t, x)$. The remaining steps are almost identical to the data-driven inference. The partial differential equation is represented by the gradients of the solution u and parameter λ with respect to the input parameters x and t. These are used to compute the mean squared error loss of the partial differential equation MSE_f, boundary conditions MSE_b and initial conditions MSE_0. The sum of the mean squared error losses is then used as a cost function as in Eq. (5.2). If the boundary or initial conditions are independent of the parameter λ, their corresponding loss terms can be neglected. Finally, the cost function is minimized by the procedure previously explained in Sect. 5.2.

5.3.1 Static Model

To illustrate the data-driven identification using a physics-informed neural network, the example presented in Sect. 5.2.1 is revisited. However, now it is assumed that the displacement $u(x)$ is known while the cross-sectional properties $EA(x)$ are to be determined. As before, the domain $\Omega = [0, 1]$ is investigated. To increase the complexity of the problem, a cubic variation in the cross-sectional properties is defined as

$$EA(x) = x^3 - x^2 + 1. \tag{5.50}$$

Again, the manufactured solution $u(x)$ is chosen as

$$u(x) = \sin(2\pi x), \tag{5.51}$$

which leads to the following distributed load $p(x)$ after insertion into the differential equation (5.4)

$$p(x) = -2(3x^2 - 2x)\pi \cos(2\pi x) + 4(x^3 - x2 + 1)\pi^2 \sin(2\pi x). \tag{5.52}$$

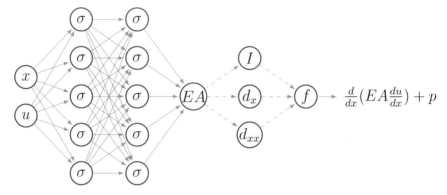

Fig. 5.10 Conceptual physics-informed neural network for data-driven identification of the static bar equation. The left part shows the feed-forward neural network and the right part represents the physics-informed neural network. The dashed lines denote non-trainable weights

The chosen solution leads to the boundary conditions

$$u(0) = u(1) = 0. \tag{5.53}$$

However, the boundary conditions do not have any influence on the learning of the model parameters, as the model parameters are not part of the boundary conditions. Therefore, the boundary loss MSE_b can be omitted.

The network architecture to solve this inverse problem, illustrated in Fig. 5.10 is similar to the one presented in Sect. 5.2.1, Fig. 5.1. The difference lies in the inputs being the coordinates x as well as the displacements u. Additionally, the left feed-forward neural network predicts the stiffness EA instead of the displacements u. The differential equation loss is identical to Eq. (5.13), while the boundary loss is omitted.

Given this information, an inverse physics-informed neural network can be constructed to predict the cross-sectional properties $EA(x)$ of the bar. This is shown in Fig. 5.11. Here, it is seen that the varying cross-section is approximated well. Additionally, the training history is shown along with the cost function.

Note that variations of this inverse network exist. Raissi et al. [RPK19] extend a neural network with a trainable parameter EA. Their neural network uses coordinates x as inputs and displacements u as output, similar to the forward-driven approach. The cost function is then defined as the sum of the mean squared error between the measurements $\hat{u}$ and the predictions u and the physics-informed loss. Tartakovsky et al. [Tar+18] use two neural networks. The first network predicts the displacement u from the coordinates x, and the second network predicts the stiffness EA from the predicted displacements u.

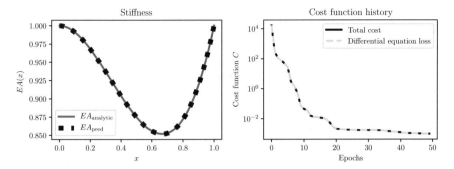

Fig. 5.11 Linear elastic static bar example. Left: the cross-sectional properties $EA(x)$ computed with a physics-informed neural network and the analytic expression are compared. Right: the training history is illustrated by showing the cost function for each epoch

5.4 Related Work

In earlier work [RPK17a, RPK17b], Gaussian processes were used for the inference and identification of differential equations. However, certain limitations imposed by the nature of Gaussian processes [RK18, RPK17c] led to the development of neural-network-based approaches [RPK19].

The introduction of physics-informed neural networks inspired fellow researchers to adapt and extend the proposed method. For instance, Pang et al. introduced a hybrid approach, that combines numerical discretization and physics-informed neural networks to solve space-time fractional advection–diffusion equations [PLK18]. Another extension is tailored to solve three-dimensional fluid flow problems by encapsulating the governing physics of the Navier–Stokes equation [RYK18]. Following classical numerical approximations from the family of Galerkin methods, Kharazmi et al. proposed a physics-enriched network, that is trained by minimizing the variational formulation of the underlying partial differential equation [KZK19]. Choosing a variational residual as the cost function reduces the order of the differential operator, and thus promises to simplify the optimization problem during training. Nevertheless, the use of a variational description does not come without limitations as the choice of integration points and the enforcement of Dirichlet boundaries require special treatment.

Since using the variational formulation of a problem is the typical approach for the numerical solution of partial differential equations, it is not surprising to find this idea also in other related publications. For example, Samaniego et al. followed this paradigm and translated the variational energy formulation of mechanical systems into the cost function of a deep learning model [Sam+19]. Next to a description of their physics-informed neural network implementation, the publication entailed several example applications from the field of computational solid mechanics, like phase-field modeling of fracture or bending of a Kirchhoff plate.

Motivated by the successful application of machine learning algorithms to the solution of high-order non-linear partial differential equations [BEJ19], E and Yu employed a deep residual neural network for solving variational problems [EY17]. Residual neural networks form an extension of feed-forward neural networks (cf. Sect. 3.1) which introduce additional connections between non-adjacent layers. In this way, some outputs can skip intermediate layers, which reduces the difficulty of training extremely deep networks [He+15]. Nabian and Meidani et al. also made use of this network architecture in order to build a surrogate model for high-dimensional random partial differential equations [NM19]. On the academic examples of diffusion and heat conduction they demonstrated that their implementation is able to deal with the strong and the variational formulation of the problem. The challenge of solving high-dimensional partial differential equations also motivated Sirignano and Spiliopoulos to investigate the applicability of deep learning in this context [SS18]. Their method can solve free-boundary partial differential equations in up to 200 dimensions. Apart from a special treatment of the free boundaries, they proposed a Monte Carlo-based method for the computation of second derivatives in higher dimensions.

The aforementioned papers dealt only with the inference of solutions to partial differential equations. An alternative approach for data-driven discovery of partial differential equations was presented by Rudy et al. [Rud+17]. They introduced a technique for discovering governing equations and physical laws by observing time-series measurements in the spatial domain. Their method allows the use of either an Eulerian or a Lagrangian reference frame.

A major drawback of the approaches introduced in the previous sections is that training a neural network is generally much more computationally expensive than applying a conventional numerical method. Since the physics-informed neural networks are learning the solution of a problem for specific initial conditions, boundary conditions and material parameters, a small change in one of them demands a complete re-training. Hence, they do not represent a viable alternative in terms of computational costs. In recent publications, Zhu et al. [Zhu+19], and Geneva and Zabaras [GZ20] tried to overcome this issue by building physics-constrained surrogate models that are able to accurately predict solutions for a range of initial conditions. Their goal was to construct a model that provides a time-discretized solution given only the initial state of the system. For that, they used an autoencoder (auto-regressive dense encoder-decoder) to produce a mapping from one time step to the other similar to a discrete time-stepping scheme. The autoencoder employed convolutional neural networks (cf. Sect. 3.10.1) that offer an effective alternative to fully connected feed-forward architectures when modeling transient partial differential equations [Zhu+19]. To train the network parameters, a cost function compares the autoencoder prediction with the result of a time integration step. Figure 5.12 shows the training procedure where the autoencoder predicts multiple time steps before the network parameters are updated with backpropagation. In contrast to physics-informed neural networks and related approaches, their method is not dependent on labeled training data, e.g. a prescribed solution at the boundaries. The initial training of the autoencoder network is still in a timely manner. However, after the training is

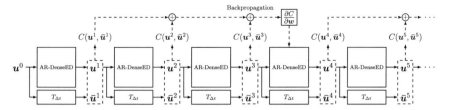

Fig. 5.12 Composition of cost function used for training the autoencoder (AR-DenseED) with backpropagation. Here, w denotes the learnable parameters and u^n represents the prediction at time step n. Further, $\bar{u}^n$ describes the target computed with the numerical time integration scheme $T_{\Delta t}$, while $C(u^n, \bar{u}^n)$ is the physics-constrained cost function. In the shown example, three predictions are considered during the parameter update with backpropagation. Image adapted from Geneva and Zabaras [GZ20] with permission from Elsevier

completed the model can be used to predict the solution of non-linear partial differential equations for a wide range of initial conditions. Geneva and Zabaras showed that in case of non-linear dynamical systems their approach yielded results magnitudes faster than classical FEM or FDM solvers. In particular, the authors were able to solve and model the Kuramoto-Sivashinsky equation as well as the one- and two-dimensional Burgers' equations. Furthermore, their method achieved satisfactory generalization properties outside the training domain. Another interesting feature introduced is a Bayesian framework that allows to quantify the uncertainty of the predicted quantities. Even though promising results were presented, their approach still exhibited certain downsides. Since a discretization in time was used, the method is prone to instabilities typical for any kind of numerical approximation. According to the authors, those issues could be resolved with the same techniques used to stabilize classical discrete time-stepping schemes.

References

[PU92]   Dimitris C. Psichogios and Lyle H. Ungar. "A hybrid neural network-first principles approach to process modeling". In: *AIChE J.* 38.10 (Oct. 1992), pp. 1499–1511. ISSN: 0001-1541, 1547-5905. DOI https://doi.org/10.1002/aic.690381003 (visited on 07/02/2020).

[LLF98]   I.E. Lagaris, A. Likas, and D.I. Fotiadis. "Artificial neural networks for solving ordinary and partial differential equations". In: *IEEE Trans. Neural Netw.* 9.5 (Sept. 1998), pp. 987–1000. ISSN: 10459227. DOI https://doi.org/10.1109/72.712178 (visited on 01/08/2020).

[Kon18]   Risi Kondor. "N-body Networks: a Covariant Hierarchical Neural Network Architecture for Learning Atomic Potentials". In: arXiv:1803.01588 [cs] (Mar. 5, 2018) (visited on 07/15/2020).

[HMP17]   Matthew Hirn, Stéphane Mallat, and Nicolas Poilvert. "Wavelet Scattering Regression of Quantum Chemical Energies". In: *Multiscale Model. Simul.* 15.2 (Jan. 2017), pp. 827–863. ISSN: 1540-3459, 1540-3467. DOI https://doi.org/10.1137/16M1075454. arXiv:1605.04654 (visited on 07/15/2020).

[Mal16]   Stéphane Mallat. "Understanding deep convolutional networks". In: *Phil. Trans. R. Soc.
             A* 374.2065 (Apr. 13, 2016), p. 20150203. ISSN: 1364-503X, 1471-2962. DOI https://
             doi.org/10.1098/rsta.2015.0203 (visited on 07/15/2020).
[RPK19]   M. Raissi, P. Perdikaris, and G.E. Karniadakis. "Physics-informed neural networks:
             A deep learning framework for solving forward and inverse problems involving
             nonlinear partial differential equations". In: *Journal of Computational Physics* 378
             (Feb. 2019), pp. 686–707. ISSN: 00219991. DOI https://doi.org/10.1016/j.jcp.2018.10.
             045. URL: https://linkinghub.elsevier.com/retrieve/pii/S0021999118307125 (visited on
             01/08/2020).
[Rai20]   Maziar Raissi. *maziarraissi/PINNs*. original-date: 2018-01-21T04:04:32Z. July 25,
             2020. URL: https://github.com/maziarraissi/PINNs (visited on 07/27/2020).
[BNK20]   Steven Brunton, Bernd Noack, and Petros Koumoutsakos. "Machine Learning for Fluid
             Mechanics". In: *Annu. Rev. Fluid Mech.* 52.1 (Jan. 5, 2020), pp. 477–508. ISSN: 0066-
             4189, 1545-4479. DOI https://doi.org/10.1146/annurev-fluid-010719-060214. arXiv:
             1905.11075 (visited on 06/26/2020).
[FDC20]   Michael Frank, Dimitris Drikakis, and Vassilis Charissis. "Machine-Learning Methods
             for Computational Science and Engineering". In: *Computation* 8.1 (Mar. 3, 2020), p.
             15. ISSN: 2079-3197. DOI https://doi.org/10.3390/computation8010015. URL: https://
             www.mdpi.com/2079-3197/8/1/15 (visited on 07/02/2020).
[Sam+19]   Esteban Samaniego et al. "An Energy Approach to the Solution of Partial Differential
             Equations in Computational Mechanics via Machine Learning: Concepts, Implementa-
             tion and Applications". In: arXiv:1908.10407 [cs, math, stat] (Sept. 2, 2019) (visited on
             01/08/2020).
[LN89]   Dong C. Liu and Jorge Nocedal. "On the limited memory BFGS method for large
             scale optimization". In: *Mathematical Programming* 45.1 (Aug. 1989), pp. 503–528.
             ISSN: 0025-5610, 1436-4646. DOI https://doi.org/10.1007/BF01589116 (visited on
             07/13/2020).
[Kol+18]   S. Kollmannsberger et al. "A hierarchical computational model for moving thermal
             loads and phase changes with applications to selective laser melting". In: *Computers &
             Mathematics with Applications* 75.5 (Mar. 2018), pp. 1483–1497. ISSN: 08981221.
             DOI https://doi.org/10.1016/j.camwa.2017.11.014. URL: https://linkinghub.elsevier.
             com/retrieve/pii/S0898122117307289 (visited on 07/21/2020).
[Kol+19]   Stefan Kollmannsberger et al. "Accurate Prediction of Melt Pool Shapes in Laser Pow-
             der Bed Fusion by the Non-Linear Temperature Equation Including Phase Changes:
             Model validity: isotropic versus anisotropic conductivity to capture AM Benchmark
             Test AMB2018-02". In: *Integr Mater Manuf Innov* 8.2 (June 2019), pp. 167–177. ISSN:
             2193-9764, 2193-9772. DOI https://doi.org/10.1007/s40192-019-00132-9 (visited on
             07/20/2020).
[Roa02]   Patrick J. Roache. "Code Verification by the Method of Manufactured Solutions". In:
             J. Fluids Eng 124.1 (Mar. 1, 2002). Publisher: American Society of Mechanical Engi-
             neers Digital Collection, pp. 4–10. ISSN: 0098-2202. DOI https://doi.org/10.1115/1.
             1436090. URL: https://asmedigitalcollection.asme.org/fluidsengineering/article/124/1/
             4/462791/Code-Verification-by-the-Method-of-Manufactured (visited on 07/23/2020).
[Ste87]   Michael Stein. "Large Sample Properties of Simulations Using Latin Hypercube Sam-
             pling". In: *Technometrics* 29.2 (May 1987), pp. 143–151. ISSN: 0040-1706, 1537-2723.
             DOI https://doi.org/10.1080/00401706.1987.10488205 (visited on 07/13/2020).
[KB17]   Diederik P. Kingma and Jimmy Ba. "Adam: A Method for Stochastic Optimization". In:
             arXiv:1412.6980 [cs] (Jan. 29, 2017) (visited on 07/30/2020).
[NM19]   Mohammad Amin Nabian and Hadi Meidani. "A Deep Neural Network Surrogate for
             High-Dimensional Random Partial Differential Equations". In: *Probabilistic Engineer-
             ing Mechanics* 57 (July 2019), pp. 14–25. ISSN: 02668920. DOI https://doi.org/10.
             1016/j.probengmech.2019.05.001. arXiv:1806.02957 (visited on 02/21/2020).

[Ise08]   Arieh Iserles. *A First Course in the Numerical Analysis of Differential Equations*. Google-Books-ID: 3acgAwAAQBAJ. Cambridge University Press, Nov. 27, 2008. 481 pp. ISBN: 978-1-139-47376-7.

[Tar+18]   Alexandre M. Tartakovsky et al. "Learning Parameters and Constitutive Relationships with Physics Informed Deep Neural Networks". In: (Aug. 2018). URL: https://arxiv.org/pdf/1808.03398.pdf.

[RPK17a]   Maziar Raissi, Paris Perdikaris, and George Em Karniadakis. "Inferring solutions of differential equations using noisy multi-fidelity data". In: *Journal of Computational Physics* 335 (Apr. 2017), pp. 736–746. ISSN: 00219991. DOI https://doi.org/10.1016/j.jcp.2017.01.060. arXiv:1607.04805 (visited on 07/16/2020).

[RPK17b]   Maziar Raissi, Paris Perdikaris, and George Em Karniadakis. "Machine learning of linear differential equations using Gaussian processes". In: *Journal of Computational Physics* 348 (Nov. 1, 2017), pp. 683–693. ISSN: 0021-9991. DOI https://doi.org/10.1016/j.jcp.2017.07.050. URL: http://www.sciencedirect.com/science/article/pii/S0021999117305582 (visited on 07/16/2020).

[RK18]   Maziar Raissi and George Em Karniadakis. "Hidden Physics Models: Machine Learning of Nonlinear Partial Differential Equations". In: *Journal of Computational Physics* 357 (Mar. 2018), pp. 125–141. ISSN: 00219991. DOI https://doi.org/10.1016/j.jcp.2017.11.039. arXiv:1708.00588 (visited on 07/16/2020).

[RPK17c]   Maziar Raissi, Paris Perdikaris, and George Em Karniadakis. "Numerical Gaussian Processes for Time-dependent and Non-linear Partial Differential Equations". In: arXiv:1703.10230 [cs, math, stat] (Mar. 29, 2017) (visited on 07/16/2020).

[PLK18]   Guofei Pang, Lu Lu, and George Em Karniadakis. "fPINNs: Fractional Physics-Informed Neural Networks". In: arXiv:1811.08967 [physics] (Nov. 19, 2018) (visited on 07/16/2020).

[RYK18]   Maziar Raissi, Alireza Yazdani, and George Em Karniadakis. "Hidden Fluid Mechanics: A Navier-Stokes Informed Deep Learning Framework for Assimilating Flow Visualization Data". In: arXiv:1808.04327 [physics, stat] (Aug. 13, 2018) (visited on 04/09/2020).

[KZK19]   E. Kharazmi, Z. Zhang, and G. E. Karniadakis. "Variational Physics- Informed Neural Networks For Solving Partial Differential Equations". In: arXiv:1912.00873 [physics, stat] (Nov. 27, 2019) (visited on 07/16/2020).

[BEJ19]   Christian Beck, Weinan E, and Arnulf Jentzen. "Machine learning approximation algorithms for high-dimensional fully nonlinear partial differential equations and second-order backward stochastic differential equations". In: *J Nonlinear Sci* 29.4 (Aug. 2019), pp. 1563–1619. ISSN: 0938-8974, 1432-1467. DOI https://doi.org/10.1007/s00332-018-9525-3. arXiv:1709.05963 (visited on 07/16/2020).

[EY17]   Weinan E and Bing Yu. "The Deep Ritz method: A deep learning-based numerical algorithm for solving variational problems". In: arXiv:1710.00211 [cs, stat] (Sept. 30, 2017) (visited on 01/14/2020).

[He+15]   Kaiming He et al. "Deep Residual Learning for Image Recognition". In: arXiv:1512.03385 [cs] (Dec. 10, 2015) (visited on 07/16/2020).

[SS18]   Justin Sirignano and Konstantinos Spiliopoulos. "DGM: A deep learning algorithm for solving partial differential equations". In: *Journal of Computational Physics* 375 (Dec. 2018), pp. 1339–1364. ISSN: 00219991. DOI https://doi.org/10.1016/j.jcp.2018.08.029. arXiv:1708.07469 (visited on 01/08/2020).

[Rud+17]   Samuel H. Rudy et al. "Data-driven discovery of partial differential equations". In: *Sci. Adv.* 3.4 (Apr. 2017), e1602614. ISSN: 2375-2548. DOI https://doi.org/10.1126/sciadv.1602614 (visited on 01/08/2020).

[Zhu+19]   Yinhao Zhu et al. "Physics-Constrained Deep Learning for High-dimensional Surrogate Modeling and Uncertainty Quantification without Labeled Data". In: *Journal of Computational Physics* 394 (Oct. 2019), pp. 56–81. ISSN: 00219991. DOI https://doi.org/10.1016/j.jcp.2019.05.024. arXiv:1901.06314 (visited on 07/06/2020).

[GZ20]   Nicholas Geneva and Nicholas Zabaras. "Modeling the Dynamics of PDE Systems with Physics-Constrained Deep Auto-Regressive Networks". In: *Journal of Computational Physics* 403 (Feb. 2020), p. 109056. ISSN: 00219991. DOI https://doi.org/10.1016/j.jcp.2019.109056. arXiv:1906.05747 (visited on 11/09/2020).

Chapter 6
Deep Energy Method

An alternative method to the data-driven inference with the physics-informed neural networks for physical systems is the deep energy method. This approach was first presented by Nguyen-Thanh et al. in [NZR19]. Instead of using the partial differential equations and minimizing the corresponding residual, the potential energy of the system is minimized. However, this is only valid for physical systems that fulfill the principle of minimum potential energy, such as, for example, structural mechanical problems under conservative loads. Furthermore, Nguyen-Tanh et al. have only discussed static problems, thus the temporal dimension t is neglected in the following discussions.

The basic idea of using a neural network to approximate the solution $u(x)$ of a partial differential equation is the same as for the physics-informed neural network. The boundary conditions can again either be enforced by a strong approach or by inclusion in the loss function. The Neumann boundary conditions are, however, often already part of the potential energy. Furthermore, a strong enforcement of the boundary conditions leads to an unconstrained optimization problem instead of a constrained optimization problem, as solely the potential energy has to be minimized. This makes it easier to estimate the solution $u(x)$ accurately.

The potential energy Π_{tot} can be expressed by the sum of the internal Π_i and the external energy Π_e

$$\Pi_{\text{tot}} = \Pi_i + \Pi_e. \tag{6.1}$$

The internal energy is computed by an integral of the internal energy density over the domain, which in structural mechanics is the product of the strains and stresses. The external energy is given by the negative work performed by the load. The external

---

Electronic supplementary material The online version of this chapter (https://doi.org/10.1007/978-3-030-76587-3_6) contains supplementary material, which is available to authorized users.

S. Kollmannsberger et al., *Deep Learning in Computational Mechanics*, Studies in Computational Intelligence 977, https://doi.org/10.1007/978-3-030-76587-3_6

energy is defined as negative, as bringing the system back to its original configuration including the applied load requires positive work. Both energy components require the computation of an integral that depends on the solution $u(x)$. As the solution is only known at specific collocation points, a numerical integration scheme has to be applied. One of the simplest choices is the trapezoidal rule, described in textbooks like [CK12], where the domain is split into trapezoids, which makes it possible to approximate the area and thereby the integral. The integration scheme can be expressed as

$$\int_a^b f(x)dx \approx \frac{\Delta x}{2}(f(x_0) + 2f(x_1) + 2f(x_2) + \cdots + 2f(x_{n-1}) + f(x_n)),$$
(6.2)

where $\Delta x = \frac{b-a}{n}$ is the distance between the integration points $x_i = a + i\Delta x$. An alternative, but very similar numerical integration scheme is the Simpson's rule, also described in [CK12], which computes the integral by splitting the domain into sub-domains and approximating the integrand with quadratic polynomials. The scheme is given as

$$\int_a^b f(x)dx \approx \frac{\Delta x}{3}(f(x_0) + 4f(x_1) + 2f(x_2) + 4f(x_3) + \cdots + 4f(x_{n-1}) + f(x_n)), \quad (6.3)$$

where Δx and x_i are calculated in the same way as for the trapezoidal rule. Simpson's rule is more accurate, but adds complexity. A third alternative is the midpoint rule, where the domain of the integral is split up into rectangles. Therefore, the integration scheme is the simplest and can be expressed as

$$\int_a^b f(x)dx \approx \sum_{i=0}^n f\left(\frac{x_{i+1} - x_i}{2}\right)\Delta x,$$
(6.4)

where Δx and x_i are computed as before. This rule is the most inaccurate, but very useful, as the function is only evaluated at the midpoints and not the boundaries. This is useful if an integrand has a singularity at the boundary, which then can be avoided in the evaluation of the integral. A summary of the three integration schemes is given in Table 6.1.

The evaluation of the integral slightly increases the complexity of the computations and makes the optimization problem more challenging. It is therefore recommended by [NZR19] to use a strong enforcement of the boundary conditions so that the optimization problem is unconstrained.

Another drawback of the deep energy method is that there is no inbuilt regularization, as it was empirically observed for the physics-informed neural network in [RPK19]. Overfitting can occur, causing unphysical oscillations of the solution $u(x)$. These oscillations lead to an excessive and unphysical external energy Π_e, which may grow toward $-\infty$, while not increasing the internal energy Π_i significantly. The computed internal energy is not increasing, as it is computed by the same set of

Table 6.1 Comparison of three numerical integration schemes. The distance between the integration points is given as $\Delta x = \frac{b-a}{n}$ and the integration points are given as $x_i = a + i\Delta x$

	Order of integration	Integration scheme
Midpoint rule	0	$\sum_{i=0}^{n} f(\frac{x_{i+1}-x_i}{2})\Delta x$
Trapezoidal rule	1	$\frac{\Delta x}{2}(f(x_0) + 2f(x_1) + 2f(x_2) + \cdots + 2f(x_{n-1}) + f(x_n))$
Simpson's rule	2	$\frac{\Delta x}{3}(f(x_0) + 4f(x_1) + 2f(x_2) + 4f(x_3) + \cdots + f(x_n))$

collocation points. If a different set of collocation points is used, an enormous internal energy would be observed and one would discover, that the potential energy was not successfully minimized. To avoid overfitting with this method, one could use the early stopping regularization, as it directly counteracts this issue. Here, a different set of collocation points is used to test- if the potential energy also decreases for this set of points or only for the training set. If overfitting is encountered, the training is stopped early and the best solution $u(x)$ for the testing set is used.

The deep energy method also has advantages. For example, it handles cases with singularities in the distributed load $p(x)$ better. This is because the singularity never has to be explicitly evaluated thanks to the numerical integration. For the physics-informed neural network these cases pose a problem, where the singularity either has to be approximated or avoided, which can decrease significantly the quality of the solution. Moreover, the deep energy method naturally avoids the evaluation of the second derivative of the network w.r.t. the input variable and instead computes the integral of the solution. The evaluation of that integral turns out to be computationally much less demanding as the computation of the second derivative, at least for the investigated examples.

6.1 Static Model

The one-dimensional static bar is very suitable to illustrate how the deep energy method works. The approach can thereby also be easily compared to the one-dimensional static bar implemented with the physics-informed neural network in Sect. 5.2.1. The simplest version of the static bar uses a linear elastic formulation. The goal is to estimate the longitudinal displacement $u(x)$ under a certain distributed load $p(x)$ and concentrated loads F_i. The internal energy is given as

$$\Pi_i = \int_\Omega \frac{1}{2} E A \left(\frac{du}{dx}\right)^2 dx, \tag{6.5}$$

where E is Young's modulus and A is the cross-sectional area of the bar. The external energy is defined as

$$\Pi_e = -\int_\Omega p(x)u(x)dx - \sum_i F_i u(x_i). \tag{6.6}$$

The potential energy is then computed as the sum of the internal and external energy in Eq. (6.1) that has to be minimal for the correct displacement $u(x)$. To compute the internal and external energy, either the trapezoidal, Simpson's or midpoint rule may be used. For the static bar, a strong enforcement of the boundary conditions is very easy. Therefore, the cost function is calculated directly with the potential energy

$$C = \Pi_{\text{tot}} = \Pi_i + \Pi_e, \tag{6.7}$$

where the Neumann boundary conditions, in form of the concentrated forces, are already part of the external energy. The Dirichlet boundary conditions are applied to the prediction $z(x)$ produced by the neural network. This is achieved using a multiplication with a function $A(x)$ and a summation with a function $B(x)$

$$u(x) = a(x)z(x) + b(x). \tag{6.8}$$

The functions $a(x)$ and (bx) are chosen such that the Dirichlet boundary conditions are fulfilled. For homogeneous boundary conditions $b(x) = 0$ and $a(x)$ is chosen correspondingly. For a clamped bar on both sides and a length of L the Dirichlet boundary conditions are given as

$$u(0) = 0, \tag{6.9}$$
$$u(L) = 0. \tag{6.10}$$

A possible choice for the strong application of the boundary conditions is $a(x) = (x - L)x$ and $b(x) = 0$. The choice is however not unique.

To illustrate the deep energy method even further, a specific example is now presented. The same manufactured solution as for the physics-informed static bar example in Sect. 5.2.1 is chosen as

$$u(x) = \sin(2\pi x), \tag{6.11}$$

for $x \in \Omega = [0, 1]$ and $L = 1$. The corresponding distributed load is found by insertion into the differential equation

$$p(x) = 4\pi^2 \sin(2\pi x) \tag{6.12}$$

with the cross-sectional properties being $EA = 1$. Using the boundary conditions, Eqs. (6.9) and (6.10) lead to a displacement that can be expressed via the prediction $z(x)$ of the neural network as

$$u(x) = (x - L)xz(x). \tag{6.13}$$

The deep energy method uses the same `buildModel` and `getDerivative` functions as described in Sect. 5.2.1 to create the neural network and calculate the derivatives. The displacements are found with the neural network and the modification due to the strong enforcement.

```
def getDisplacement(x):
    z = model(x)
    u = (x - 1) * x * z
    return u
```

Furthermore, a numerical integration scheme has to be implemented. Here, the trapezoidal rule is chosen, where it is assumed that the points are distributed uniformly.

```
def trapezoidalIntegration(y, x):
    dx = x[1] - x[0]
    result = torch.sum(y)
    result = result - (y[0] + y[-1]) / 2
    return result * dx
```

Finally, the internal energy Π_i and external energy Π_e can be computed with the following two functions.

```
def getInternalEnergy(u, x, EA):
    e = getDerivative(u,x)
    internal_energy_density = 0.5 * EA(x) * e ** 2
    internal_energy = trapezoidalIntegration(internal_energy_density, x)
    return internal_energy

def getExternalEnergy(u, x):
    external_energy_density = p(x) * u
    external_energy = trapezoidalIntegration(external_energy_density, x)
    return external_energy
```

Everything can then be combined to calculate the potential energy Π_{tot} for the presented example.

```
model = buildModel(1, 10, 1)
x = torch.linspace(0, 1, 10, requires_grad=True).view(-1, 1)
EA = lambda x: 1 + 0 * x
p = lambda x: 4 * math.pi**2 *torch.sin(2 * math.pi * x)

u = getDisplacement(x)
internal_energy = getInternalEnergy(u, x, EA)
external_energy = getExternalEnergy(u, x)
potential_energy = internal_energy + external_energy
```

Now it is possible to minimize the cost function C, which is the potential energy Π_{tot}, via a stochastic gradient descent or more sophisticated schemes like the L-BFGS optimizer. After the minimization it is possible to make an accurate prediction of the displacement $u(x)$. Algorithm 7 summarizes how the deep energy method applied to a static bar works. When comparing it to Algorithm 4 for the physics-informed neural network, the only difference is the cost function.

Algorithm 7 Training the deep energy method for the static solution of the problem described in Eq. (5.4).

Require: training data for boundary condition $\{x_b^i, u_b^i\}_{i=1}^{N_b}$

    generate N_f collocation points with a uniform distribution $\{x_f^i\}_{i=1}^{N_f}$
    define network architecture (input, output, hidden layers, hidden neurons)
    initialize network parameters $\boldsymbol{\Theta}$: weights $\{\boldsymbol{W}^l\}_{l=1}^{L}$ and biases $\{\boldsymbol{b}^l\}_{l=1}^{L}$ for all layers L
    set hyperparameters for L-BFGS optimizer ($epochs$, learning rate $\alpha, \dots$)

    for all $epochs$ **do**
        predict $\hat{\boldsymbol{u}} \leftarrow \boldsymbol{u}_{NN}(\boldsymbol{x}; \boldsymbol{\Theta})$
        $\frac{\partial \hat{\boldsymbol{u}}_b}{\partial x} \leftarrow \frac{\partial \boldsymbol{u}_{NN}}{\partial x}(\boldsymbol{x}_b; \boldsymbol{\Theta})$
        $\boldsymbol{f} \leftarrow \boldsymbol{f}_{NN}(\boldsymbol{x}_f; \boldsymbol{\Theta})$
        compute Π_i, Π_e using numerical integration   ▷ cf. Eqs. (6.5) and (6.6)
        compute cost function: $C \leftarrow \Pi = \Pi_i + \Pi_e$
        update parameters: $\boldsymbol{\Theta} \leftarrow \boldsymbol{\Theta} - \alpha \frac{\partial C}{\partial \boldsymbol{\Theta}}$ ▷ L-BFGS
    end for

The computed displacements and the analytical solution for the discussed example are compared in Fig. 6.1. One observes that the estimation of the solution $u(x)$ is very accurate, as no significant differences can be noticed. Additionally, Fig. 6.1 shows the training history. Here, one can see the different energy terms over the number of epochs of training. It can also be seen that the found potential energy is negative. This is due to the way the external energy was defined. Clapeyron's theorem for linear elasticity [FT03] states that the potential energy is equal to half of the external work. This is reflected in Fig. 6.2b.

The overfitting issue with the deep energy method can be illustrated with the same example by increasing the number of training epochs without introducing any regularization to the neural network. This is shown in Fig. 6.2. It is easy to spot the point at which overfitting occurs. The energy values start diverging from there on, where the external and potential energy go toward $-\infty$. The overfitting can also be seen by looking at the displacements of the bar which have large errors now. As stated before, overfitting can be avoided by introducing early stopping.

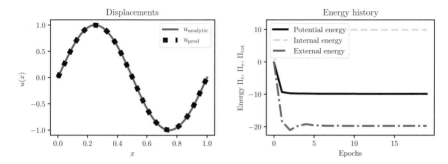

Fig. 6.1 Linear elastic static bar example. Left: the displacements $u(x)$ computed with the deep energy method and the analytic solution are compared. Right: the training history is illustrated by showing the cost function, i.e. the potential energy for each epoch

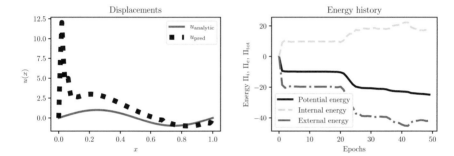

Fig. 6.2 Linear elastic static bar example. Left: the displacements $u(x)$ computed with the deep energy method and the analytic solution are compared. Right: the training history is illustrated by showing the cost function, i.e. the potential energy for each epoch. Overfitting occurs after 22 epochs

References

[CK12]   Ward E. Cheney and David R. Kincaid. *Numerical Mathematics and Computing*. 7th. Cengage Learning, 2012. 704 pp. ISBN: 978-1-133-10371-4.

[NZR19] Vien Minh Nguyen-Thanh, Xiaoying Zhuang, and Timon Rabczuk. "A deep energy method for finite deformation hyperelasticity". In: *European Journal of Mechanics - A/Solids* (Oct. 2019), p. 103874. ISSN: 09977538. DOI https://doi.org/10.1016/j.euromechsol.2019.103874. URL: https://linkinghub.elsevier.com/retrieve/pii/S0997753819305352 (visited on 01/08/2020).

[RPK19] M. Raissi, P. Perdikaris, and G.E. Karniadakis. "Physics-informed neural networks: A deep learning framework for solving forward and inverse problems involving nonlinear partial differential equations". In: *Journal of Computational Physics* 378 (Feb. 2019), pp. 686–707. ISSN: 00219991. DOI https://doi.org/10.1016/j.jcp.2018.10.045. URL: https://linkinghub.elsevier.com/retrieve/pii/S0021999118307125 (visited on 01/08/2020).

[FT03]   Roger Fosdick and Lev Truskinovsky. "About Clapeyron's theorem in linear elasticity". In: *Journal of Elasticity* 72 (Mar. 2003), pp. 145–172. URL: http://www.cityu.edu.hk/ma/ws2007/notes/FTJElast.pdf.

Appendix A
Exercises

Exercises for Chap. 2

Exercise 1—Linear Regression

1. Find the cost function as defined in Eq. (2.3) using a linear regression with an initialization of the weight $w = 1$ and bias $b = 1$. The following data is to be used.

$$x = [0, 2, 3]$$
$$y = [1, 6, 7]$$

2. Find the gradient of the cost function with respect to the weight and bias.
3. Find the minimum of the cost function and the corresponding weight and bias directly using the gradients. Then plot the linear regression.
4. Now repeat the linear regression with the matrix approach and compare your results with the linear regression from the previous exercise.
5. Find the minimum of the cost function and the corresponding weight and bias by using a gradient descent approach. Use the initial weight of $w = 1$, the initial bias $b = 1$, and a learning rate of $\alpha = 0.1$. Do 3 iterations and check if you converge toward the result of the previous exercise.
6. Now the goal is to implement a linear regression class in python. A template for the class `LinearRegression.py` is provided. The correctness of the implemented class can be tested using the unit-test defined in `LinearRegression_Test.py`. The first task is to implement the member function `predict (self, x)`, which takes an input x and returns the linear prediction.
7. The next member function to be implemented is `costFunction(self, x, y)`. Here, the cost function is to be found with the input arrays x and y. The output should be a scalar value.

© The Editor(s) (if applicable) and The Author(s), under exclusive license to Springer Nature Switzerland AG 2021
S. Kollmannsberger et al. *Deep Learning in Computational Mechanics*,
Studies in Computational Intelligence 997, https://doi.org/10.1007/978-3-030-76587-3

8. Now the member function `gradient(self, x, y)` has to be added. Here, the inputs x and y are again arrays that have to be used to determine the scalar gradients of the cost function with respect to the weight and bias. The output is defined as an array of the gradient with respect to the bias and the gradient with respect to the weight.

9. Finally, the member function `train(self, epochs, learningRate)` has to be implemented. This function approximates the weight and bias that minimize the cost function by using the gradient descent approach. The number of iterations is provided as an input `epochs` and the learning rate is provided as `learningRate`. The function has to update the weight and bias after each iteration. Additionally, the value of the cost function using the training and testing data can be printed on screen. To reduce the number of outputs for the user, this should only be done for every 10th or 100th iteration.

10. Now the class can be applied to an example case defined in `example2_1.py`. Here, a linear regression is applied to random data with a linear tendency. You can try to play around with the sample size, number of epochs, and learning rate.

Exercise 2—Higher-Order Regression

1. The goal is now to extend the linear regression class to a higher-order regression class. The class is very similar with almost the same member functions. Only slight adjustments are necessary. An outline of the class is provided as `HigherOrderRegression.py` and can be tested with `HigherOrder Regression_Test.py`. The first step is to implement the member function `predict(self, x)`. Here, a higher-order prediction has to be made using the following form,

$$y = \sum_{i=1}^{p} (w_i x^i) + b$$

where p is the polynomial degree, that is selected during construction of the class. One thing to note here is that the weight w_i is an array instead of a scalar now.

2. The member function `costFunction(self, x, y)` has to be adjusted accordingly.

3. Now derive the gradients of the cost function with respect to all weights and the bias. The derived quantities can then be inserted in the member function `gradient(self, x, y)`. The function should return an array with all the gradients with respect to the weights and with respect to the bias. The gradient with respect to the bias should be the last entry.

4. Finally, adjust the member function `train(self, epochs, learning Rate`, such that you ensure that all the weights are adjusted accordingly using the gradient descent approach.

5. Now, the class can be used in an example case defined in `example2_2.py`. Try to adjust the model parameters, such as the polynomial degree, the number of epochs, and the learning rate. Which phenomena occur, when a polynomial

degree of 1, 2, and 9 is chosen? How can this phenomenon already be spotted during learning and how can it be avoided?

6. To avoid overfitting, regularization is now to be added to the higher-order regression class. If the parameter `regularization` is set to `True`, an L^2 regularization has to be used. Adjust the member functions `costFunction` and `gradient` accordingly.

7. Apply the regularization to `example2_2.py` and see if the model is still overfitting. Then try to vary the regularization parameter. What happens, when the parameter goes toward zero, and what happens when it goes toward infinity?

Exercises for Chap. 3

Exercise 1—Approximating the Sine Function

This exercise revisits the example of a simple feed-forward neural network approximating the sine function which was introduced in the manuscript in Sect. 3.8. The goal of this exercise is to deepen the understanding of neural networks by showing the influence of different hyperparameters.

1. Change the neural network architecture to a single hidden layer with two neurons and run the script `exercise3.py`. What do you observe?
2. Now set the network to two hidden layers with 50 neurons each. What phenomena can be observed?
3. Name ways to overcome this problem.
4. Uncomment `Model.plotTrainingHistory()` and rerun the code to plot the cost function history. At which epoch should the algorithm stop to prevent the model from overfitting?
5. Set the learning rate to $\alpha = 0.1$. What do you observe?
6. Now reset the learning rate to $\alpha = 0.001$ and apply L^2-regularization by setting the value for $\lambda > 0$. Find a value where the model fits the sine curve without overfitting.
7. Increase the regularization parameter λ until underfitting occurs and plot the result.
8. Change the model to use a rectified linear unit (ReLU) activation function and plot the result.
   Hint: Take a look at the function `buildModel` in function `approximator.py`.

Exercise 2—Neural Network Derivatives

1. Express the output $\hat{y}$ of the neural network depicted in Fig. A.1 as a function of the inputs x_1, x_2 and the network parameters, the weights

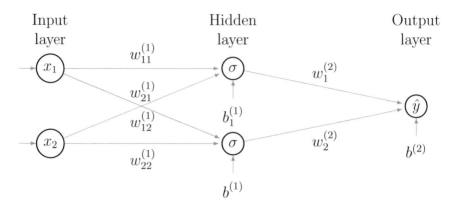

Fig. A.1 A simple feed-forward neural network example

$$\boldsymbol{W}^{(1)} = \begin{bmatrix} w_{11}^{(1)} & w_{12}^{(1)} \\ w_{21}^{(1)} & w_{22}^{(1)} \end{bmatrix}, \quad \boldsymbol{w}^{(2)} = \begin{bmatrix} w_1^{(2)} \\ w_2^{(2)} \end{bmatrix}$$

and biases

$$\boldsymbol{b}^{(1)} = \begin{bmatrix} b_1^{(1)} \\ b_2^{(1)} \end{bmatrix}, \quad b^{(2)}$$

2. Compute all partial derivatives $\frac{\partial \hat{y}}{\partial w_{jk}^{(l)}}$ and $\frac{\partial \hat{y}}{\partial b_j^{(l)}}$ assuming the activation function σ is given as $\sigma = (\cdot)^2$.
3. The cost function is defined as $C = (y - \hat{y})^2$. Compute the partial derivatives of the cost function C w.r.t. the weights and biases.

Hint: $\frac{\partial C}{\partial w_{jk}^{(l)}} = \frac{\partial C}{\partial \hat{y}} \frac{\partial \hat{y}}{\partial w_{jk}^{(l)}}$

Exercises for Chap. 5

Exercise 1—Physics Informed Neural Network for a Static Bar

A bar in the domain $x \in [0, \frac{3}{2}]$ is given with the varying cross-sectional properties

$$EA(x) = x^2 + 1.$$

The goal is to produce a manufactured solution to later test the implementation of a physics-informed neural network. The following displacement solution is prescribed

$$u(x) = 1 - \cos(3\pi x).$$

1. Compute the distributed load $p(x)$ from the differential equation given in the manuscript. Additionally, state the two homogeneous boundary conditions.
2. Now the goal is to implement a physics-informed neural network with PyTorch for a static bar. For this purpose, a class `PhysicsInformedBarModel` will be implemented which can be found in the `PhysicsInformed.py` file. Additionally, some helper functions are required. These are to be implemented in the `utilities.py` file. One can check the correctness of the implementation with the unit tests in the `PhysicsInformed_Test.py` file. Finally, an example driver file `exercise5_1.py` is provided such that the network can be applied to an example case.

   First the function `generateGrid1d(length, samples=20, initial_coordinate=0.0)` has to be implemented. The function should return a uniformly spaced grid starting at the initial coordinate with a total length given by the length parameter and with the chosen number of samples. The grid should be returned as a column tensor with the shape (samples, 1). Additionally, you have to ensure that `requires_grad=True` is selected inside the tensor sampling over the domain so the gradients can be computed at a later stage.
   Hint: Two useful commands for this exercise are `torch.linspace` and `torch.Tensor.view`.
3. The next step is to implement the member function `costFunction(self, x, u_pred)`. Here, one has to return the value of the differential equation cost and the value of the boundary condition cost separately. The cost functions are to be determined as discussed in the manuscript in Eq. (5.13) for the differential equation and in equation Eq. (5.12) for the boundary conditions. The boundary conditions are given by a list of a list in the following manner.
   [[value, order of derivative, index],[value, order of derivative, index],...].
4. Run the `exercise5_1.py` file and try changing the parameters. What happens if you increase or decrease the learning rate, number of epochs or the weighting factor for the boundary conditions?
5. Now try to reproduce the example from the manuscript in Sect. 5.2.1 by replacing the parameters in the `exercise5_1.py` file. Try to find good model parameters to estimate the solution as fast and precise as possible.

Exercise 2—Data-Driven Identification Using Physics-Informed Neural Networks for a Static Bar

You are given the following information on a static bar defined in the domain $x \in [0, 1]$

$$EA(x) = 1 - u(x),$$
$$p(x) = \sin(x),$$
$$u(x) = 1 - \sqrt{2\sin(1)x - 2\sin(x) + 1}.$$

1. What is an important difference between the given data and the examples provided in the manuscript?
2. The goal is now to implement an inverse physics-informed neural network for a static bar. This is to be done in the `PhysicsInformed.py` file, where the structure of the class is already provided. The same `utilities.py` from before is to be used. Additionally, an example file `exercise5_2.py` is implemented with the data provided above.
   Start by implementing the member function `predict(self, x, u)` that predicts the parameter $EA(x)$ from the coordinate x and displacement u.
   Hint: A useful command is `torch.cat`.
3. Implement the two member functions
   `costFunction(self, x, u, EA_pred)` and `train(self, x, u epochs, **kwargs)`. The member function `costFunctionself, x, u, EA_pred` only returns the differential equation cost, as the boundary terms are neglected. If you get stuck, take a look at Exercise 5.1 and check how these member functions were implemented in the forward version of the physics-informed neural network.
4. It is now possible to run the example file `exercise5_2.py`. Try playing around with the model parameters to get an intuition for their influence.
5. Try to reproduce the example from the manuscript in Sect. 5.3 by replacing the parameters in the `exercise5_2.py` file. Try to find good model parameters to estimate the solution as fast and precise as possible.

Exercise 5.3—Normalization and Weighting

1. The influence of the scaling of the solution u and the weighting of the different loss terms, the physics and boundary terms, are investigated. For this, the example bar from Exercise 5.1 is revisited. Further, the file `exercise5_3.py` is provided. The file is almost identical to the file provided in Exercise 5.1. The difference lies in the scaling factor `scale` which scales the solution u and is initially set to 1. At the same time, the differential equation is adjusted by dividing Young's modulus E by the scaling factor. The goal is to investigate the influence of this scaling factor. Try changing the factor to 10^2 and then 10^{-2}. What do you observe? If the given training parameters inhibit convergence of the minimization procedure, change them. You can adjust the number of epochs, the learning rate, or change the optimizer. Additionally, you can even try to use multiple training blocks, for example, first using an Adam optimizer and then a L-BFGS optimizer. What can you conclude for the scaled cases?
2. Now a weighting factor is introduced on the boundary condition loss, `bc_wei ght`. Initially, it is set to 1. Try varying this weighting factor, while keeping the remaining hyperparameters constant. You can track the total cost after 500

epochs and fill out the following table:

Total cost	bc_weight
	10^{-4}
	10^{-3}
	10^{-2}
	10^{-1}
	10^{0}
	10^{1}
	10^{2}
	10^{3}
	10^{4}

What do you observe?

Hint: Note that results might vary, due to the random initialization. It is best to run the cases multiple times and average the results.

Exercises for Chap. 6

Exercise 1—Deep Energy Method of a Static Bar

1. A bar defined in the domain $x \in [0, 1]$ is given with the cross-sectional properties $EA = 1$ and the displacement

$$u(x) = -x^{0.65} + 0.65x.$$

   Now derive the distributed load $p(x)$ that has to be applied for the given displacement. Additionally, identify from the displacement $u(x)$ the Dirichlet boundary condition at $x = 0$, and the Neumann boundary condition at $x = 1$.
2. Try to use the Physics Informed Neural Network implemented in exercise 5 to estimate the displacement $u(x)$. What is the issue and how can it be avoided?
3. Compute the internal Π_i, external Π_e and potential energy Π_{tot} as defined in the manuscript analytically.
4. Repeat the computation of the energy terms using the Midpoint rule. The integrands have already been evaluated at the midpoints for $\Delta x = 0.2$ and are given in the following table.

x_i	0.1	0.3	0.5	0.7	0.9
$d\Pi_i$	0.324	0.0580	0.0159	0.00373	0.000298
$d\Pi_e$	−0.809	−0.303	−0.181	−0.124	−0.0915

Is the numerical integration close to the analytical expressions derived in the previous exercise? If not, explain why it isn't and how it can be improved.

5. The goal is now to implement the Deep Energy Method with PyTorch for a static bar and see if it is possible to solve the given problem without compromises. The file `utilities.py` includes the helper functions, the file `DeepEnergy.py` includes the definition of the class `DeepEnergyBarModel`, and the driver file is given as `exercise6_1.py`. Again unit tests are supplied via the file `DeepEnergy_Test.py`. The first step is to implement the two integration functions `trapezoidalIntegration1d(y, x)` and `midpointInteg ration1d(y, x)` using the corresponding integration schemes provided in the manuscript.
   Hint: Assume, that Δx is constant throughout the domain.

6. Next, implement the member functions of the class `DeepEnergyBarModel`. Start with the `getDisplacements(self, x)` function. Here, you have to use the neural network to predict a function $z(x)$ and then enforce the Dirichlet boundary conditions strongly. There are three options for the Dirichlet boundary conditions, `'fixed_left'`, `'fixed_right'`, and `'fixed_both'`.

7. Now implement the member function `getStrains(self, u, x)`.

8. Finally the three member functions for the energy calculations are to be implemented. These are the `getInternalEnergy(self, u, x)`, `getExtern alEnergy(self, u, x)`, and `getEnergyValues(self, u, x)`. The last energy function returns all three energy terms, the internal energy, external energy, and the potential energy.

9. Now run the example file `exercise6_1.py` using the Midpoint integration and then the Trapezoidal integration. Why does only one of them work?

10. Compare the computed energy terms to the expected analytical energy terms. Why is there a difference?

11. Comment on the accuracy of the predicted displacement. Is it possible to increase the number of epochs to improve accuracy? Try to change the optimizer from L-BFGS to Adam and comment on the result.

12. Now try to reproduce the example from the manuscript in Sect. 6 by replacing the parameters in the `exercise6_1.py` file. Try to find good model parameters to estimate the solution as fast and precise as possible.

13. What happens when you increase the number of epochs and how could this phenomenon be avoided?

Appendix B
Additional Figures

B.1 Section 5.2.2

(See Fig. B.1)

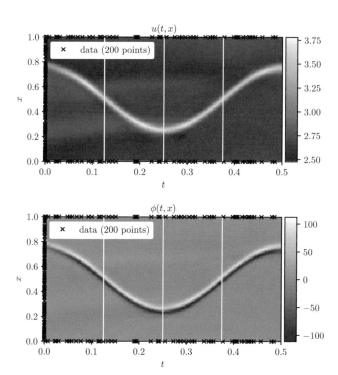

Fig. B.1 Prediction of the temperature distribution u and corresponding heat flux ϕ for an unbalanced cost function. Top: Approximated solution and location of time snapshots (white lines). Bottom: Comparison of predicted and exact solution at distinct snapshots

© The Editor(s) (if applicable) and The Author(s), under exclusive license to Springer
Nature Switzerland AG 2021
S. Kollmannsberger et al. *Deep Learning in Computational Mechanics*,
Studies in Computational Intelligence 997, https://doi.org/10.1007/978-3-030-76587-3

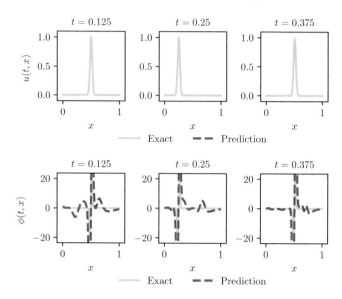

Fig. B.1 (continued)

Index

S. Kollmannsberger et al. *Deep Learning in Computational Mechanics*,
Studies in Computational Intelligence 997, https://doi.org/10.1007/978-3-030-76587-3

Printed in the United States
by Baker & Taylor Publisher Services